Recent Trends in Waste Water Treatment and Water Resource Management

Sadhan Kumar Ghosh · Papita Das Saha ·
Maria Francesco Di
Editors

Recent Trends in Waste Water Treatment and Water Resource Management

Springer

Editors
Sadhan Kumar Ghosh
Department of Mechanical Engineering
Jadavpur University
Kolkata, West Bengal, India

Papita Das Saha
Department of Chemical Engineering
Jadavpur University
Kolkata, West Bengal, India

Maria Francesco Di
Università degli Studi di Perugia
Perugia, Italy

ISBN 978-981-15-0708-3   ISBN 978-981-15-0706-9   (eBook)
https://doi.org/10.1007/978-981-15-0706-9

This Springer imprint is published by the registered company Springer Nature Singapore Pte Ltd.
The registered company address is: 152 Beach Road, #21-01/04 Gateway East, Singapore 189721, Singapore

Preface: Everything Flows, but Circularity is the Solution

There is nothing more essential to life on Earth than water. Yet, there is a global water crisis. People are struggling to access the quantity and quality of water they need for drinking, cooking, bathing, hand washing and growing their food. Rapid industrialization, population growth and unplanned urbanization have contributed to severe land and water pollution. The main sources of freshwater pollution can be attributed to direct discharge of untreated toxic industrial wastes and dumping of various industrial effluents. More than 40% of the world's population resulted affected by water scarcity, and this problem is also featured in the UN Sustainable Development Goal 6 (SDG 6), "clean water and sanitation". Currently, about 3 in 10 people lack safe drinking water and 6 in 10 lack inadequate sanitation systems. Globally, 844 million people lack access to clean water resulting in families and communities being locked in poverty for generations. Access to clean water as a stepping stone to development changes everything. Safe, secure, affordable and equitable access to clean water in particular for the poor is among the most relevant challenge to be faced in the next years. In 1993, the UN General Assembly designates March 22 as World Water Day and recognizes the importance of addressing the global water crisis each year with an amazing progress in making clean drinking water accessible to 2.6 billion people in developing countries from 1990 to 2015. There are still opportunities to multiply the benefits of clean water through improved sanitation and hygiene education.

The wastewater contains contaminants such as metal ions and organic compounds impacting the environment severely. This aspect will be further exacerbated in the next years by the increase in urbanization, in particular in developing areas, leading to the use of more of Earth's dwindling resources. In recent years, nanotechnology has introduced a myriad of novel nanomaterial that can have promising outcomes in environmental clean-up and remediation. Thus, it is highly needed now to recycle and reuse water resources in a sustainable manner. Effluent discharged in water bodies contributes to eutrophication, pathogen growth and emission of nitrous oxide and methane with GHG potential of about 300 and 21 times higher than CO_2, respectively. Untreated effluents have impacts three times higher than the treated ones representing a key factor of paramount importance for safe and environmentally sustainable urbanization. Among the multiple pollutants like fluorides, heavy metals, dyes, oils, etc., several new emerging components, such as endocrine-disrupting compounds, pharmaceuticals, personal care products and other bioactive organic compounds, are also present in the wastewaters.

The conventional wastewater treatment processes such as ion exchange, electrocoagulation, membrane filtration and adsorption are used. Adsorbents from renewable resources like tamarind seed, agricultural waste materials, Bermuda grass, rice husk and tea wastes are used nowadays. Carbonaceous materials are being used widely nowadays, which include hydrochar, biochar and activated carbon. These adsorbents are environmentally sustainable and available from waste products like MSW and domestic wastes. Various low-cost sorbents including waste/by-products from industries and agricultural operations, natural materials and microbial and non-microbial biomass have been identified and documented as effective dye-removing agents. Synthesizing cellulose nanofibres from agricultural lignocellulosic wastes is a novel approach for the production of a material from a cheap renewable resource. Recent development of membrane separation technologies, nanomaterial-based waste separation processes and advanced oxidation processes like combinations of ozone, UV light and H_2O_2 could be the alternative to these problems. Private and public sectors should invest in infrastructure to create the conditions for the transition to a circular use of resources bringing about social, economic and environmental benefits. Implementation of circular economy in this sector, including the safe reuse of water and the energy neutrality, can have also positive consequences on water quality and water security.

The 8th IconSWM 2018 received 380 abstracts and 320 full papers from 30 countries. A total of 300 accepted full papers have been presented in November 2018 at Acharya Nagarjuna University, Guntur, Andhra Pradesh, India. After thorough review by experts and required revisions, the board has finally selected 25 chapters in this book, "Recent Trends in Waste Water Treatment and Water Resource Management", dealing with the application of synthesized nanocellulose material, defluoridation of water, removal of hexavalent chromium, methylene blue dye, lead, chloride and different iodine species, sustainable sewage treatment process, utility of bacteria for detoxification of textile dyes, utilization of activated carbon, anaerobic co-digestion of sewage sludge and animal by-product and many modern techniques of wastewater treatment.

The IconSWM movement was initiated focusing better waste management, resource circulation and environmental protection since the year 2009. It helps in generating awareness and bringing all the stakeholders together from all over the world under the aegis of the International Society of Waste Management, Air and Water (ISWMAW). It established a few research projects across the world involving the CST at Indian Institute of Science, Jadavpur University, and a few other institutions in India and experts from more than 30 countries in the research project on circular economy. Consortium of Researchers in International Collaboration (CRIC) and many other organizations across the world help the

IconSWM movement. IconSWM has become one of the biggest platforms in India for knowledge sharing and awareness generation among the Urban Local Bodies (ULBs), government departments, researchers, industries, NGOs, communities and other stakeholders in the area of waste management. The primary agenda of this conference is to reduce the waste generation encouraging the implementation of 5Rs (Reduce, Reuse, Recycle, Repair and Remanufacturing) concept. The conference provided holistic pathways to waste management and resource circulation conforming to urban mining and circular economy.

The success of the 8th IconSWM is the result of effective contribution of the government of Andhra Pradesh, several industry associations, chamber of commerce and industries, the AP Higher Education Council, and various organizations and individuals in India and abroad. Support of the UNEP, UNIDO and UNCRD and delegation from European Union and other foreign organizations was significant. 8th IconSWM 2018 was attended by nearly 823 delegates from 22 countries. The 9th IconSWM 2019 will be held at KIIT, Bhubaneswar, Odisha, during 27–30 November 2019, expecting nearly 900 participants from 30 countries.

This book will be helpful for the educational and research institutes, policy-makers, government implementers, ULBs and NGOs. We hope to see you all in 9th IconSWM 2019 in November 2019.

July 2019

Prof. Sadhan Kumar Ghosh
Jadavpur University
Kolkata, India

Prof. Papita Das Saha
Jadavpur University
Kolkata, India

Prof. Maria Francesco Di
Università degli Studi di Perugia
Perugia, Italy

Acknowledgements

We thank the Hon'ble Chief Minister and Hon'ble Minister of MA&UD for taking personal interest in this conference.

We are indebted to Shri. R. Valavan Karikal, IAS; Dr. C. L. Venkata Rao; Shri. B. S. S. Prasad, IFS (Retd.); Prof. S. Vijaya Raju and Prof. A. Rajendra Prasad, VC, ANU, for their unconditional support and guidance for preparing the platform for successful 8th IconSWM at Guntur, Vijayawada, AP.

I must express my gratitude to Mr. Vinod Kumar Jindal, ICoAS; Shri. D. Muralidhar Reddy, IAS; Shri. K. Kanna Babu, IAS; Mr. Vivek Jadav, IAS; Mr. Anjum Parwez, IAS; Mr. Bala Kishore; Prof. S. Varadarajan; Mr. K. Vinayakam; Prof. Shinichi Sakai, Kyoto University, JSMCWM; Prof. Y. C. Seo and Prof. S. W. Rhee of KSWM; Shri. C. R. C. Mohanty of UNCRD; members of Industry Associations in Andhra Pradesh; Prof. P. Agamuthu, WM&R; Prof. M. Nelles, Rostock University; Dr. Rene Van Berkel, UNIDO; Ms. Kakuko Nagatani-Yoshida and Mr. Atul Bagai of UNEP and UN Delegation to India for their active support.

IconSWM-ISWMAW Committee acknowledges the contribution and interest of all the sponsors, industry partners, industries, co-organizers, organizing partners around the world, the government of Andhra Pradesh, Swachh Andhra Corporation as the principal collaborator, the vice chancellor and all the professors and academic community at Acharya Nagarjuna University (ANU), the chairman, vice chairman, secretary and other officers of AP State Council of Higher Education for involving all the universities in the state, the chairman, member secretary and the officers of the AP Pollution Control Board, the Director of Factories, the Director of Boilers, the Director of Mines and officers of different ports in Andhra Pradesh and the delegates and service providers for making a successful 8th IconSWM.

I must specially mention the support and guidance of each of the members of the International Scientific Committee, CRIC members, the core group members and the local organizing committee members of 8th IconSWM who are the pillars for the success of the programme. The editorial board members including the reviewers, authors and speakers and Mr Aninda Bose and Ms Kamiya Khatter of

M/s. Springer India Pvt. Ltd deserve thanks who were very enthusiastic in giving me inputs to bring this book.

I must mention the active participation of all the team members in IconSWM Secretariat across the country with special mention of Prof. H. N. Chanakya and his team in IISc Bangalore; Ms. Sheetal Singh and Dr. Sandhya Jaykumar and their team in CMAK and BBMP; Mr. Saikesh Paruchuri; Mr. Anjaneyulu; Ms. Senophiah Mary; Mr. Rahul Baidya; Ms. Ipsita Saha; Mr. Suresh Mondal; Mr. Bisweswar Ghosh; Mr. Gobinda Debnath and the research team members in the Department of Mechanical Engineering and ISWMAW, Kolkata HQ, for various activities for the success of the 8th IconSWM 2018.

I express my special thanks to Sannidhya Kumar Ghosh, being the governing body member of ISWMAW supported the activities from USA. I am indebted to Mrs. Pranati Ghosh who gave me guidance and moral support in achieving the success of the event. Once again, the IconSWM and ISWMAW express gratitude to all the stakeholders, delegates and speakers who are the part of the success of 8th IconSWM 2018.

Contents

About the Editors

Prof. Sadhan Kumar Ghosh, Ph.D., is the Dean, Faculty of Engineering and Technology and Professor & Former Head of the Mechanical Engineering Department, as well as the Founder Coordinator of the Centre for QMS at Jadavpur University, India. A prominent researcher in the fields of Waste Management, Circular Economy, SME Sustainability, Green Manufacturing, Green Factories and TQM, he has served as the Director, CBWE, Ministry of Labour and Employment, Government of India and L&T Ltd. Prof Ghosh is the Founder and Chairman of the IconSWM and the President of the International Society of Waste Management, Air and Water. He is the India PI of the JU in India of the Horizon 2020 Project "India H2O", project funded by European Union (2018–2022) concerning wastewater to drinking water in India. He was awarded a Distinguished Visiting Fellowship by the Royal Academy of Engineering, UK, to work on "Energy Recovery from Municipal Solid Waste" in 2012. He received the Boston Pledge and NABC 2006 Award for the most eco-friendly innovation "Conversion of plastics & jute waste to wealth" in the ESP/50K Business Plan Competition in Houston, Texas, USA. In addition, he holds patents on waste plastic processing technology and high-speed jute ribboning technology for preventing water wastage and occupational hazards.

Prof. Papita Das Saha, Ph.D., is a Professor of Chemical Engineering, Jadavpur University, Kolkata, India. She holds a B.Sc. in Chemistry; B.Tech. in Chemical Engineering; and an M.E. and Ph.D. in Chemical Engineering. She has 12 years of teaching experience and has published more than 100 papers in international journals. She is also a committee member of the West Bengal Pollution Control Board, Department of West Bengal, and serves as an editor/regional editor for various international journals.

Prof. Maria Francesco Di, Ph.D., is currently an Associated Professor at the Dipartimento di Ingegneria of the University of Perugia. To date, he has coordinated more than 30 projects financed by private and public companies/institutions and the European Commission. He is the Head of the LAR5 Laboratory (Reuse, Recycle, Recovery of Waste and Residues) at the Dipartimento di Ingegneria of the University of Perugia. He was and currently is involved in a Technical Working Group of the European Commission focusing on waste management and green procurements. His research interests, as reflected in more than 190 publications, include renewable energy production, integrated waste and wastewater management, disposal, recycling and recovery of waste and residues, with a particular focus on integrated, sustainable approaches.

Anaerobic Co-digestion of Sewage Sludge and Animal by-Product

Ahmad Reza Salehiyoun, Maria Francesco Di, Mohammad Sharifi, Omid Noroozi, Hamid Zilouei and Mortaza Aghbashlo

Abstract The management of animal by-products (ABP) is a crucial issue since the relevant economic and environmental consequences that this practice involves. Anaerobic digestion (AD) is indicated as a suitable process for addressing such issue. AD facilities resulted already largely implemented in existing wastewater treatment plants for sludge stabilization but operated a quite low organic loading rate (kgVS/m^3). For this reason, thence digestion of ABP with SS in existing facilities can be considered a solution to a double problem: efficient treatment of ABP at reduced investment costs; increase in energy efficiency of existing digesters. In this study, the effect of different ABP to sewage sludge ratios on AD performances was investigated by batch runs. Best results were achieved for a ratio of 40% w/w leading to a biomethane yield of 730 L/kgVS with a VS reduction of 70%. No significant inhibition phenomena occurred during the runs.

Keywords Anaerobic digestion · Animal by-product · Co-digestion · Sewage sludge · Biomethane potential

A. R. Salehiyoun · M. Sharifi · M. Aghbashlo
Department of Mechanical Engineering of Agricultural Machinery, Faculty of Agricultural Engineering and Technology, College of Agriculture and Natural Resources, University of Tehran, Karaj, Iran

M. Francesco Di (✉) · O. Noroozi
LAR5 Laboratory, Department of Engineering, University of Perugia, Via G. Duranti, 06125 Perugia, Italy
e-mail: francesco.dimaria@unipg.it

H. Zilouei
Department of Chemical Engineering, Isfahan University of Technology, 84156-83111 Isfahan, Iran

© Springer Nature Singapore Pte Ltd. 2020
S. K. Ghosh et al. (eds.), *Recent Trends in Waste Water Treatment and Water Resource Management*, https://doi.org/10.1007/978-981-15-0706-9_1

1 Introduction

Due to the high nutritional value of meat, meat consumption has increased in the world and its production has exacerbated centralized activities, causing large animal by-product (ABP) in slaughterhouses (Moukazis et al. 2018). Approximately, between 20 and 50% of weight of a live animal is not edible for human (Pagés-Díaz et al. 2014).

In the only EU area, the amount of deadweight of animals generated by slaughterhouses is of about 39 M tonnes. Of these 6.3 M tonnes are represented by bovine and 17.5 M tonnes by swine. At the Italian level, the total amount of dead weight of animals processed in slaughterhouses is of about 3.77 M tonnes of which 1.1 M tonnes are bovine and 1.4 M tonnes swine. According to European Union legislation for slaughterhouse by-products management (EC 2009), animal waste is divided into three categories, namely Category 1, 2 and 3, based on the reduced risk of utilization. Majority of the slaughterhouse waste (SHW) belongs to Category 2 and 3 of ABP, which can be used in various ways under certain conditions, provided that not be hazardous to animal and human health.

Different methods for management of ABP have been proposed including incineration (for energy), aerobic digestion (composting), alkaline hydrolysis and anaerobic digestion (AD) (Moukazis et al. 2018). Anaerobic digestion is one of the best options for slaughterhouse waste treatment which produce energy-rich biogas, reducing greenhouse gases emissions and effectively control the pollution in abattoirs. Category 2 and 3 can be used as a feedstock to AD after completing a series of pretreatment including crushing, pressure sterilization (heating in 133 °C, 3 bar for 20 min) and pasteurization (size reduction below 6–12 mm and heating to 70 °C for 60 min) (Ortner et al. 2014).

AD of solid and pasty slaughterhouse waste fractions is much more difficult, and in contrast to wastewater, only a few industrial applications are established (Ortner et al. 2014). Anaerobic digestion of solid slaughterhouse waste is engaged several problems owning to its high nitrogen and sulfur concentration. The anaerobic treatment of these wastes as mono-substrates often leads to the accumulation of ammonia, volatile fatty acids (VFA) and long chain fatty acids (LCFA). These may inhibit methanogenesis and thereby lead to instability, lower biogas production and, at higher concentrations, lead to complete inhibition (Di Maria et al. 2016; Hejnfelt and Angelidaki 2009; Pitk et al. 2013). Dissolved free ammonia (NH_3) and sulfides are most responsible for process imbalances, causes, e.g., reduced methane yields and the inhibitory effects of intermediate substances (Chen et al. 2008). For successful operation of anaerobic digestion at high concentrations of ammonia, measures such as adjustment of C/N ratio by adding high-carbon substrates, acclimatization of microflora to higher ammonia concentrations, control and modification of pH, reduction in ammonia concentration by stripping or membrane technique, and two-stage anaerobic digestion have been proposed (Rajagopal et al. 2013). The most problems often occurred in non-appropriate C/N ratio. Many studies proposed in the literature investigated the possibility of increasing the performance of these existing digesters by co-digestion with other substrates. Co-digestion can supply macro- and micro-nutrient, vary in bacteria consortium, balance C/N, improve buffer capacity and process stability, dilute inhibitors and supply water resource for process and increment biogas production with respect to mono-digestion (Karthikeyan and Visvanathan 2013).

A relevant sector in which AD is currently fully implemented is the wastewater treatment plant (WWTP). In this sector in the only EU, there are about 36,000 WWTPs. Despite the development of many AD treatment plants in Italy, most of the designed anaerobic digesters have exceeded the required capacity, so organic loading rates in their digesters are low, which results in the digester to operate in an inefficient and non-economic way. For this reason, in the present work the possibility of co-digestion of ABP with sewage sludge for an existing Italian WWTP was investigated.

In this research, the effect of adding the total stream of slaughterhouse waste originated from a small-scale bovine-pig abattoir in Italy to the sewage sludge (SS) was investigated in different proportions in batch assays. The results of this research will contribute to assess the sustainability of the exploitation and management of ABP by an integrated approach strongly related to human bean basic needs as food and wastewater.

2 Materials and Methods

2.1 Slaughterhouse Waste and Inoculum

The slaughterhouse waste was prepared from a pig and bovine slaughterhouse in the city of Marsciano, Italy, which slaughtered 15,000 pigs, 2000 cows and a limited number of sheep per week. The bovine slaughterhouse waste was selected and collected in all sections of categories 2 and 3 according to the European standard (EC 2009). Each section was first grinded with industrial meat grinder up to 6 mm and then was homogenized in a large blender (Fimar, Italy) and was finally stored in a freezer at −20 °C until it was consumed. Prior to consumption, the pasteurization operation was performed at 70 °C for 1 h in an oven in a sealed container.

Municipal sewage sludge was prepared from a WWTP of 90,000 population equivalent in the town of Foligno (Italy). The sludge sample was sampled after a concentrator including primary and waste activated sludge. Sewage sludge was stored at 4 °C. The inoculum was sampled from the anaerobic digester of the WWTP and acclimatized at 37 °C for 2 weeks. When necessary the total solids (TS) concentration of the sludge was increased by settling operation.

2.2 Experimental Setup

Experiments in batch mode were carried out in 0.5-L glass digesters at mesophilic conditions (37 °C ± 1 °C) by an incubator (Fig. 1a). Different SHW to SS ratios were used: $R1 = 0\%$; $R2 = 2\%$; $R3 = 40\%$; $R4 = 60\%$ (Borowski and Kubacki 2015; Pitk et al. 2013). Experiments were performed at 4% TS. The volume of the reactor fluid was filled with 400 g of the feed and inoculum. The ratio of inoculum to feedstock was set to 1:3 on VS basis. When necessary, TS concentration was adjusted by deionized water. Before starting, each reactor was flushed with nitrogen gas for 2 min. Control samples were also activated to determine the amount of biogas produced from inoculum. Reactors were shaken manually once a day, and all the runs were activated in triplicate.

Biomethane volume was determined directly by measuring the displaced water volume (Fig. 1a) and expressed at standard pressure and temperature (273 K and 1 atm) (Kafle and Kim 2013). The percentage of methane was measured via extraction of a 10 ml sample by syringe from the reactor head and injecting and passing through a 3 molar solution to remove carbon dioxide and hydrogen sulfide (Fig. 1b).

The weighted average was used to calculate the average methane percentage (El-Mashad and Zhang 2010). The experiments were stopped when the daily biogas production reached less than 1% of cumulative one (VDI standard 2006).

Analysis of variance (ANOVA) was carried out using the SPSS18 in order to find a significant difference between the treatments in a completely randomized design (one-factor ANOVA) for biomethane yield, methane content and VS destruction. The significant difference between results was considered when p-value was lower 0.05 at a probability level of 5% ($\alpha = 0.05$).

(a) **(b)**

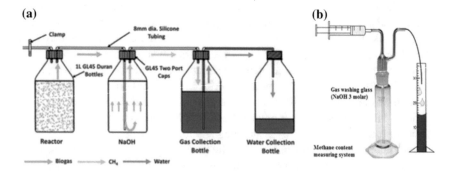

Fig. 1 Schematic of the batch laboratory-scale reactors (Ware and Power 2016) (**a**), and gas scrubber for determining of methane content (**b**)

2.3 Chemical Analysis

The total solid content was determined by placing the wet samples in an oven at 105 °C for 24 h. The dried samples were then milled and combusted to determine the VS percentage at 550 °C in a muffle furnace until it reached a constant weight (APHA 2004). The pH of the feedstock and inoculum was measured by a PH meter METTLER TOLEDO. The total amount of organic carbon (TOC) was determined in the percentage of TS according to APHA's 5310 standard (APHA 2004). The amount of total Kjeldahl nitrogen (TKN) was assessed using an autokjeldahl device and subsequent titration with 0.1 normal sulfuric acid according to ISO 1871 standard (ISO 1871: 2009 (E); ČSN, 2010). To measure the total ammonium nitrogen (TAN) and total volatile fatty acids (TVFA), the HACH Lange DR 3900 Spectrophotometer was used. The digestate samples were first centrifuged at 6000 rpm for 10 min and then filtered through 45 µ paper. Total volatile fatty acids equivalent to mg/L of acetic acid and ammonium were measured by LCK 365 kuvet and LCK 303 kuvet, respectively. The amount of VS degradation as an indicator of anaerobic degradation efficiency in converting materials and reducing its contamination load was calculated by measuring the VS of each reactor after digestion and comparing it with VS before the process.

3 Results and Discussion

3.1 Feed Characteristics

The characteristics of the SHW and municipal SS are presented in Table 1. For the former, TS and VS were 25.6% and 95.6%, respectively, very similar to those reported in the study of Hejnfelt and Angelidaki (2009), with high degradable VS. C/N ratios resulted much higher than those reported in other studies, in the range of 6–10 (Borowski and Kubacki 2015), given that total carbohydrate wastes (ruminal

Table 1 Characteristics of slaughterhouse waste and inoculum

Particular	SS		SHW		INO	
TS	1.73	0.01	25.6	0.18	2.97	0.15
TS/VS	78.6	0.17	95.6	0.04	48.5	0.93
PH	6.49		6.14		7.45	
TOC (g/kg TS)	46.9	0.27	72.7	2.18	44.0	0.75
TKN (g/kg TS)	4.00	0.22	5.81	0.22	3.78	0.01
C/N	11.7	0.51	12.5	0.49	11.7	0.14
Alkalinity ratio	–		–		0.14	

and intestinal contents) and digestive tract are considered as a mixture with the waste flow (Pitk et al. 2013). Therefore, there is a better balance in terms of preventing the ammonia inhibition.

3.2 Batch Experiments

Mean results of the batch runs are summarized in Table 2. The cumulative BMP (L/kg VS) per day, as well as the daily production of methane, is represented in Fig. 2.

Table 2 Biogas production results in a 4% TS batch experiment

Parameter	4% TS							
	0% (sewage sludge)		20% ABP		40% ABP		60% ABP	
Temperature	37		37		37		37	
OLR (g VS/L)	31.4		37.1		37.8		38.0	
Biomethane yield (L/kg VS)	434.8	45.6	717.1	29.4	736.4	9.2	674.8	15.1
CH4 content (%)	67.2	1.1	70.5	1.0	71.4	0.4	71.5	1.4
VS removal (%)	32.60	0.38	36.64	0.26	36.70	0.47	36.83	0.79
T90% (day)	16.2		12.8		13.1		12.6	

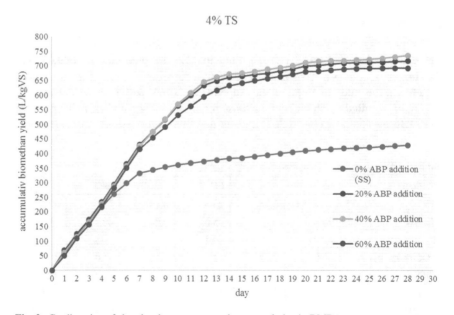

Fig. 2 Co-digestion of slaughterhouse waste and sewage sludge in BMP tests

The results of ANOVA test showed mixer had a significant effect on biomethane production ($p > 0.05$). As seen in Fig. 2, the co-digestion has doubled the production of biomethane potential indicating the presence of synergic effects. There is no significant difference between R2, R3 and R4 treatments, but the highest methane yield was obtained for R3 treatment, 736.4 L/kgVS. Anaerobic digestion of single feed wastewater has been reported in the range of 200–350 L/kgVS (Luostarinen et al. 2009; Pitk et al. 2013; Wan et al. 2011), whereas in this study achieved 470 L/kgVS, which is significantly higher. Biomethane potential for co-digestion of SHW with other substrates reported in the literature was quite in line with those of the present study. Borowski and Kubacki (2015) reported 832.9 L/kgVS for the co-digestion of SHW from pig, whereas Moukazis et al. (2018) reported values of about 855 L/kgVS. Hejnfelt and Angelidaki (2009) reported a value of 620 L/kgVS. For sterilized and cooked slaughterhouse waste (at 133 °C and 3 bar for 20 min), the best results were obtained for 5% adding to SS by 618.9 L/kgVS in semi-continuous reactor, that was equivalent to a 60% TS share and approximately 65% VS in the feedstock mixture as reported by Pitk et al. (2013). By adding 14% of grease trap sludge from a meat processing plant with 25.4% TS to SS, Luostarinen et al. (2009) reported a biomethane yield of 788 L/kgVS in a batch digestion test. The synergistic effect of ABP co-digestion with SS was superior to other organic waste supplementation, such as agricultural waste. In the Pagés-Díaz et al. study (2014), the BMP, in a 2L batch test, was obtained 613 L/kgVS for bovine slaughterhouse waste with animal fertilizer, 647 L/kgVS with the organic fraction of municipal solid waste and 461 L/kgVS with agricultural wastes (various crops, i.e., straw and fruit/vegetable waste). The relatively high nitrogen content, the high total solid (TS) content, often excludes the possibility of treating animal by-products in their original undiluted form. Therefore, dilution is typically necessary or, as a more attractive option, co-digestion with less concentrated organic waste types, such as manure or municipal wastewater (Hejnfelt and Angelidaki 2009).

Analysis of variance for methane content indicated that methane percentage was significantly affected by feed mixture. According to Table 2, an increase in the proportion of ABP has led to a significant increase in methane composition. This is due to the increased share of fat waste because lipids produce higher methane percentages than carbohydrates and proteins (Wellinger et al. 2013). However, there is no significant difference between R2, R3 and R4 treatment.

The effect of ABP:SS proportion was significant on VS degradation. The VS removal for co-digestion treatments were about 36.7%, while for SS was about 32.6%. In the AD of alkaline pre-treated mixed slaughterhouse waste, a 47% VS reduction was obtained under mesophilic conditions (Flores-Juarez et al. 2014). Although VS degradation is usually higher for AD of easily biodegradable wastes, such as food and vegetable waste, addition of slaughterhouse waste to SS reduced the effective process time (T90%) (Table 2) to an average value of about 13 days instead of the about 16 days detected for the mono-digestion of SS. The effective digestion time or technical digestion time is an important parameter that provides primary information of hydraulic retention time (HRT) for continuous anaerobic digestion (Kafle and Kim 2013). This figure resulted higher than the HRT reported

Table 3 Characteristics of digestate at the end of the batch tests

Component	4% TS			
	0%	20%	40%	60%
pH	7.27 (0.06)	7.32(0.07)	7.32 (0.10)	7.33 (0.07)
TVFA (mg/L)	141 (7)	167 (0)	167 (12)	156 (12)
TAN (mg/L)	444 (20)	564 (18)	473 (88)	587 (41)
COD (mg/L)	605 (174)	738 (230)	741 (218)	744 (219)

by Moukazis et al. (2018) that resulted in the range of 6–8 days. One of the reasons of this result can be found in the presence of fiber-containing and lignocellulosic parts such as ruminal content in the ABP composition, leading to an increase in the T90.

All the runs showed a final pH similar to the one of the mono-digestion of SS (Table 3). The final COD value for SS was only 605 mg/L. For the co-digestion runs, it ranged from 736 to 744 mg/L. The initial COD content of the sewage sludge was 2038 mg/L. Thus, an average COD removal of about 70% occurred for the runs with co-digestion of ABP and SS. Also these values resulted in line with those reported by other authors (Moukazis et al. 2018). The total amount of volatile fatty acids resulted quite low, <170 mg/L indicating the absence of specific inhibition risks. In their works, Flores-Juarez et al. (2014) and Moukazis et al. (2018) reported for AD after alkaline pretreatment of SHW a TVFA concentration of 555.5 mg/L and of 420 mg/L, respectively. No sing of ammonia inhibition was observed and TAN was in the stable spectrum, i.e., <1500–7000 mg/L (Hejnfelt and Angelidaki 2009).

4   Conclusion

Co-digestion of animal by-product and sewage sludge resulted to be an effective solution for matching two different aspects: environmentally sound management of ABP; increase the amount of energy recoverable. In particular, the large presence of wastewater treatment plant already equipped with anaerobic digester for the sludge stabilization represent an interesting opportunity also as a consequence of the reduced costs for investment, just limited to the necessary improvements for the introduction of the ABP in the existing digesters. Even at high ABP rates, the biological processes occurred with quite high efficiency and practically absence of inhibition phenomena.

References

APHA. (2004). *Standard methods for the examination of water and wastewater* (20th ed.). American Public Health Association, American Water Works Association, Water Environment Federation, Washington DC. Stand. Methods 541. ISBN 9780875532356.

Borowski, S., & Kubacki, P. (2015). Co-digestion of pig slaughterhouse waste with sewage sludge. *Waste Management, 40,* 119–126. https://doi.org/10.1016/j.wasman.2015.03.021.

Chen, Y., Cheng, J. J., & Creamer, K. S. (2008). Inhibition of anaerobic digestion process: A review. *Bioresource Technology, 99,* 4044–4064.

Di Maria, F., Micale, C., & Contini, S. (2016). Energetic and environmental sustainability of the co-digestion of sludge with bio-waste in a life cycle perspective. *Applied Energy, 171,* 67–76.

EC. (2009). Regulation (EC) No 1069/2009 of the European Parliament and of the Council of 21 October 2009 laying down health rules as regarding animal by-products and derived product not intended for human consumption and repealing Regulation (CE) No 1774/2002 (Animal by-product Regulation). Official Journal of the European Union L.300/1, 14.11.2009.

El-Mashad, H. M., & Zhang, R. (2010). Biogas production from co-digestion of dairy manure and food waste. *Bioresource Technology, 101*(11), 4021–4028.

Flores-Juarez, C. R., Rodríguez-García, A., Cárdenas-Mijangos, J., Montoya-Herrera, L., Godinez Mora-Tovar, L. A., Bustos-Bustos, E., et al. (2014). Chemically pretreating slaughterhouse solid waste to increase the efficiency of anaerobic digestion. *Journal of Bioscience and Bioengineering, 118,* 415–419. https://doi.org/10.1016/j.jbiosc.2014.03.013.

Hejnfelt, A., & Angelidaki, I. (2009). Anaerobic digestion of slaughterhouse by-products. *Biomass and Bioenergy, 33,* 1046–1054. https://doi.org/10.1016/j.biombioe.2009.03.004.

Kafle, G. K., & Kim, S. H. (2013). Anaerobic treatment of apple waste with swine manure for biogas production: Batch and continuous operation. *Applied Energy, 103,* 61–72. https://doi.org/10.1016/j.apenergy.2012.10.018.

Karthikeyan, O. P., & Visvanathan, C. (2013). Bio-energy recovery from high-solid organic substrates by dry anaerobic bio-conversion processes: A review. *Reviews in Environmental Science and Bio/Technology, 12*(3), 257–284.

Luostarinen, S., Luste, S., & Sillanpää, M. (2009). Increased biogas production at wastewater treatment plants through co-digestion of sewage sludge with grease trap sludge from a meat processing plant. *Bioresource Technology, 100,* 79–85. https://doi.org/10.1016/j.biortech.2008.06.029.

Moukazis, I., Pellera, F. M., & Gidarakos, E. (2018). Slaughterhouse by-products treatment using anaerobic digestion. *Waste Management, 71,* 652–662. https://doi.org/10.1016/j.wasman.2017.07.009.

Ortner, M., Leitzinger, K., Skupien, S., Bochmann, G., & Fuchs, W. (2014). Efficient anaerobic mono-digestion of N-rich slaughterhouse waste: Influence of ammonia, temperature and trace elements. *Bioresource Technology, 174,* 222–232. https://doi.org/10.1016/j.biortech.2014.10.023.

Pagés-Díaz, J., Pereda-Reyes, I., Taherzadeh, M. J., Sárvári-Horváth, I., & Lundin, M. (2014). Anaerobic co-digestion of solid slaughterhouse wastes with agro-residues: Synergistic and antagonistic interactions determined in batch digestion assays. *Chemical Engineering Journal, 245,* 89–98. https://doi.org/10.1016/j.cej.2014.02.008.

Pitk, P., Kaparaju, P., Palatsi, J., Affes, R., & Vilu, R. (2013). Co-digestion of sewage sludge and sterilized solid slaughterhouse waste: Methane production efficiency and process limitations. *Bioresource Technology, 134,* 227–232. https://doi.org/10.1016/j.biortech.2013.02.029.

Rajagopal, R., Massé, D. I., & Singh, G. (2013). A critical review on inhibition of anaerobic digestion process by excess ammonia. *Bioresource Technology, 143,* 632–641.

Wan, C., Zhou, Q., Fu, G., & Li, Y. (2011). Semi-continuous anaerobic co-digestion of thickened waste activated sludge and fat, oil and grease. *Waste Management, 31,* 1752–1758. https://doi.org/10.1016/j.wasman.2011.03.025.

Ware, A., & Power, N. (2016). What is the effect of mandatory pasteurisation on the biogas transformation of solid slaughterhouse wastes? *Waste Management, 48,* 503–512.

Wellinger, A., Murphy, J. D., & Baxter, D. (Eds.) (2013). *The biogas handbook: Science, production and applications.* Elsevier.

Application of Synthesized Nanocellulose Material for Removal of Malachite Green from Wastewater

Lopamudra Das, Niladri Saha, Papita Das Saha, Avijit Bhowal and Chiranjib Bhattacharya

Abstract A method associated with nanocellulose material derived from sugarcane bagasse as bio-adsorbent in order to decolorize Malachite green dye was carried out through batch process. Objective of this study was to appraise the performance of SB nanocellulose, was synthesized using bleaching treatment of sugarcane bagasse as agro-waste followed by alkaline hydrolysis, acid hydrolysis (using 50% sulfuric acid at 50 °C for 3 h), and freeze drying. The synthesized cellulose was characterized by Fourier transform infrared (FTIR) spectroscopy. The absorbance of the supernatant solution was measured at 617 nm in UV visible spectrophotometer. Adsorption process was carried out by varying the pH value (2, 4, 6, 7, 8), initial dye concentration (1, 5, 10, 15, 20, 25, 50 mg/L) contact time (15, 30, 45, 60, 120, 180 min), temperature (25, 30, 35, 40 °C), and adsorbent doses (2, 4, 6, 8 g/L). In case of SB Nano cellulose, percentage of dye removal with initial 20 mg/L dye solution was 98.35% at normal condition. The adsorption efficiency of the SB cellulose is still over 80% after four times of recycling. Adsorbent has some beneficial features like, it is renewable, inexpensive, high surface area, unique chemical composition, and top of it all these adsorbents exhibit high removal capacity in dye solution.

Keywords Adsorption · Malachite green dye · Sugarcane bagasse (SB) · SB nanocellulose

1 Introduction

With the improvement of technology, most of the countries have boosted their economy through industrialization, and by virtue of increasing industrialization, waste disposal and pollution concern become more and more important. Most industries require large amounts of water for their work. Effluents of these industries are either disposed into bare oceans or rivers, which seriously hampers our ecosystem. Dyes are widely used in the textile, cosmetics, printing, leather, paper, and electroplating industries (Hu et al. 2013). Effluents of those industries contain large amount of untreated synthetic dyes which are often difficult to biodegrade due to its xenobiotic

L. Das · N. Saha · P. D. Saha (✉) · A. Bhowal · C. Bhattacharya
Jadavpur University, Kolkata, India

© Springer Nature Singapore Pte Ltd. 2020
S. K. Ghosh et al. (eds.), *Recent Trends in Waste Water Treatment and Water Resource Management*, https://doi.org/10.1007/978-981-15-0706-9_2

properties and complex organic structures (Ma et al. 2012). The discharged dye staff wastewater from those industries can cause great threat to marine life as well as human health due to its toxicity and carcinogenic nature (Attallah et al. 2013). For example, Malachite green (MG) is one of the common types of dye in waste disposal, widely used in the textile industry. Up to now, the main conventional methods for dye removal from wastewater have been used such as membrane filtration, ion exchange, adsorption, chemical coagulation, reverse osmosis, ozonation, and ultra-filtration. (Nada and Hassan 2003; Singh et al. 2011; Silva Filho et al. 2013).

However, the adsorption technique is considered as most efficient, easy operating, inexpensive, and eco-friendly technique for wastewater treatment (Mittal et al. 2010). Carbonaceous material and nanomaterial having high surface area, high adsorption capacity, microporous structure, low cost, and special surface reactivity for the organic compound shows great efficiency in adsorption process.

Cellulose is the most ubiquitous biodegradable renewable natural polymer (Salas et al. 2014). Cellulose nanomaterials (CNs) can be isolated from various natural resources like plant, animal, or mineral plants. Two general type of cellulose nanomaterials (CNs) are commonly known as cellulose nanocrystals (CNCs) and cellulose nanofibrils (CNFs). Carbon nanocrystal particles are highly crystalline in nature (at least one dimension should be equal or less than 100 nm) (Kumar et al. 2014). It can be used as a cheaper adsorbent for removing dyes from aqueous solution directly (Ren et al. 2013; Zhou et al. 2014). Nanocellulose having high surface area, high mechanical strength, chemical inertness, and hydrophilic surface makes it a very efficient material in membranes and filters, for adsorbing impurities from industrial effluents and drinking waters. Due to intrinsic hydrophilicity property nanocellulose has organic fouling and bio-fouling reducing capacity (Tashiro and Kobayashi 1991; Šturcová et al. 2005; Mansouri et al. 2010).

India is one of the agricultural countries in the world. A large number of agro-based industries, such as jute, cotton, and sugar produce large quantities of agro-waste which hamper in proper waste disposal management and environmental ecosystem. In recent years, researchers are giving attention for questing toward maximizing the efficiency of the use of raw materials and minimizing the accumulation of waste by reusing agricultural residues. Sugarcane bagasse is fibrous matter that left after crushing the sugarcane stalks to extract the juice commonly available from the sugar and industry. Sugarcane bagasse contains about 40–55% of cellulose, hemicellulose as much as 26–35% (Wulandari et al. 2016; Mandal and Chakrabarty 2011; Jacobsen and Wyman 2002). In the present research work, sugarcane bagasse has been used as raw material for the extraction of microscopic crystal cellulose and nanocellulose obtained by delignification with sodium chlorite followed by alkylation. Interestingly, an acid hydrolysis treatment of SB has been done to get the more pure cellulose fractions (Sun et al. 2004). To generate nano-cellulose acid hydrolysis of isolated SB cellulose with 50% sulphuric acid was carried out (Wulandari et al. 2016). Characterization of nanocellulose has been made by analyzing of Fourier transform infrared spectroscopy (FTIR) (Wulandari et al. 2016). Additionally, desorption and recycling experiment provided to investigate its good cycle performance. We also discuss on the MG dye adsorption process onto SB cellulose. Thus, this study assessed the suitability of cellulose, extracted from sugarcane bagasse in removal of dye from aqueous solution.

2 Materials and Methods

2.1 Raw Bagasse

Sugarcane bagasse (SB) was obtained from the local juice center, near Jadavpur University, Kolkata, West Bengal. It was washed followed by drying in the electric oven at 120 °C for overnight. After that, the SB were ground and sieved to at range 80–100 µm. The adsorbent was stored for further use in batch adsorption.

2.2 Isolation of SB Cellulose

For each gram of raw sugarcane bagasse (SB), 50 ml of 0.75% sodium chlorite (Merck Life Science private limited) solution was used to bleach SB and it was kept for 1 h at 45 °C with continuous stirring. The treated mass was filtered, washed with distilled water until it reached to pH 5 and put into 2% sodium hydrogen sulfite (Merck Life Science private limited) solution for 30 min. It was then filtered and washed with distilled water. The dried mass was then treated with 17% NaOH (Merck Life Science private limited.) solution for 15 min to remove hemicellulose. Finally, this chemically treated mass was filtered and washed with distilled water to remove the alkali until to neutral pH.

2.3 SB Nanocellulose Preparation

Hemicellulose and lignin-free cellulose were acid hydrolyzed for 3 h at 50 °C with constant stirring. For each gram of cellulose, 20 ml of 50% H_2SO_4 (w/w) solution (Merck Life Science private limited) was used for acid hydrolyzation. After that, excess acid was washed with distilled water until it was reached to neutral pH. Hydrolysis process was followed by centrifugation and freeze drying.

2.4 Dye Solution Preparation

In this study, Malachite green (MG) dye (Loba. Chemie.) Stock solution of 1000 ppm concentration was prepared and solutions of different concentrations ranging from 1 to 50 mg/L were made by further dilutions. The absorbance of solution was measured using UV-V is spectrophotometer (Perkin Elemer—L365).

2.5 Characterization

Fourier transform infrared spectra (FTIR) of SB cellulose sample was analyzed on an instrument (PerkinElmer) in the range of 4000–450 cm^{-1} at a resolution of 4 cm^{-1}. Samples were grounded and blended with KBr. Mixture was then pressed into ultra-thin pellets. Various functional groups present on SB cellulose adsorbent have been shown by The FTIR spectrum with respect to their peak value.

2.6 Batch Adsorption Experiment

The batch adsorption study on MG dye was performed in a BOD shaker (GB Enter-prises, India) at different experimental condition such as pH (2–8), initial dye con-centration (5–25 mg/L), adsorbent mass (2–8 g/L), and temperature (25–40 °C). The pH of the dye solution was adjusted using 0.1 M NaOH (Merck Life Science private limited) and 0.1 M HCl (Merck Life Science private limited) solution. The isotherm study was done by varying the concentration of dye solution keeping temperature constant and agitated at 120 rpm. The kinetic study was investigated with differ-ent contact time. Thermodynamic study was done with varying temperature. The sample mixture solution was withdrawn from the shaker followed by centrifugation. After that, concentration of MG dye was determined by UV-VIS spectrophotometer (Perkin Elmer—L365). The dye removal efficiency of SB cellulose adsorbent and adsorbent capacity (mg/g) was derived according to the following equations:

$$\text{Removal efficiency} = \left(\frac{C_0 - C_t}{C_0} \right) \times 100$$

$$q_e = \frac{(C_0 - C_t) \times V}{m}$$

where C_0 is represented as the initial dye concentration and C_t (mg/L) is indicated as the dye concentration at time t, respectively. q_e is denoted as the amount of adsorbed dye (mg/g) per gram of adsorbent. V is referred to the volume (L) of the dye solution and m is denoted as adsorbent weight.

3 Result and Discussion

3.1 Adsorbents Analysis—FTIR

By analyzing FTIR spectroscopic spectrum of raw SB and extracted SB cellulose, different characteristics of absorption peaks were analyzed. The peak centered at 1035.57 cm^{-1} for SB is attributed to C–O stretching. Absorption band at 1728.37 cm^{-1} is attributed to the C=O stretching associated with the acetyl and uronic ester groups present in pectin and hemicellulose (Sain and Panthapulakkal 2006; Sun et al. 2005). The peaks at 1603.51 and at 1514.39 cm^{-1} are assigned to the aromatic C=C present in lignin (Garside and Wyeth 2003; Wang et al. 2009). The peak at 1242.56 cm^{-1} as present only in spectra of SB is assigned to the C–O stretching vibration of the aryl group present in lignin (Troedec et al. 2008). The broadband around 3500–3200 cm^{-1} with the maxima centered at 3332.64 cm^{-1} is attributed to the free O–H stretching present in cellulose molecules. The spectra of both samples showed C–H stretching vibration around 2898.75 cm^{-1} (Khalil et al. 2001). From spectra of raw SB, it was observed that the absorption band at 3329 cm^{-1} is attributed to the O–H stretching and that of 2899.95 cm^{-1} is attributed to the C–H stretching vibration. The peak at 1316.29 cm^{-1} is assigned to the C–H bending and C–C stretching. The peak at 1157.17 cm^{-1} arises from C–O stretching. The peak at 1026.8 cm^{-1} is attributed to the C–O–C vibration due to the presence of pyranose ring skeletal. A band at 895.61 cm^{-1} is associated with the C–O–C stretching at b-glycosidic linkage between glucose present in cellulose. Remarkably, the absence of the band at 1514.39 cm^{-1} in the spectra inferred that the prepared cellulose are free of residual lignin.

3.2 Effect of pH

As shown in Fig. 1, pH plays an important role for MG removal on SB cellulose. Initially, the pH of the dye solutions was adjusted pH range from 2 to 8 by adding HCl and NaOH solution. With increasing pH value from 2 to 8, dye adsorption efficiency

Fig. 1 Effect of pH on % removal

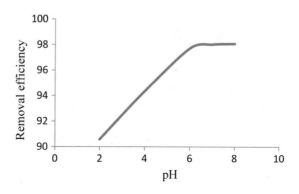

Fig. 2 Effect of contact time on % removal

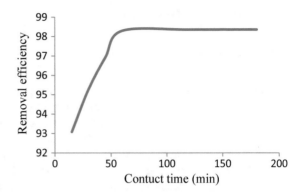

increases from 90 to 98%. Cationic dyes are favorably adsorbed on negatively charged surfaces (Saliby et al. 2013). Protonation increases in strong acidic solutions. Due to the presence of excess H^+ ion in lower pH solution and having small ionic radius, it can be adsorbed more easily than cationic MG onto SB cellulose surface. With increasing pH, more negative charged adsorption sites will be vacant. For this reason, electrostatic attraction between SB cellulose surface and cationic dye becomes easier (Yagub et al. 2014; Song et al. 2015).

3.3 Effect of Contact Time

Variation in contact time plays significant role in determining the MG dye removal efficiency of SB cellulose. Batch equilibrium studies were conducted at 30 °C by adding 8 g/L of SB cellulose adsorbent to 20 mg/L MG dye of 50 ml solution at neutral pH. The flasks were agitated in BOD shaker at 120 rpm for 2 h. Samples were taken at different time intervals (15, 30, 45, 60, 120 min) and concentration was analyzed by UV-VIS spectrophotometer. The adsorption capacity increases till the contact time increases up to 1 h and then it attained saturation condition after 2 h. It was observed that agitation time increased from 15 to 60 min, and MG dye removal efficiency of SB Cellulose increased from 93.08 to 98.35% (Fig. 2).

3.4 Effect of Adsorbent Mass

The removal efficiency was increased as the adsorbent masses were increased from 2 to 8 g/L at 50 ml of MG dye solution on equilibrium time of 2 h. It was observed that removal of MG dye changes from 88 to 98.34% with increasing adsorbent mass. From Fig. 3, it is readily estimated that by increasing the amount of adsorbent, the number of active adsorption sites on adsorbent surface was increased; therefore, dye

Fig. 3 Effect of adsorbent mass on % removal

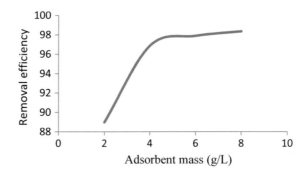

removal efficiency was increased. So, in this project work, adsorbent dose of 8 g /L was kept constant for further investigation.

3.5   *Effect of Initial Dye Concentration*

Figure 4 shows the influence of initial dye concentration on the MG dye removal efficiency by SB cellulose. The effect of initial MG dye concentration (5–50 mg/L) on removal efficiency was conducted in a BOD shaker at 120 rpm while adsorbent dose was constant at 8 g/L. It was found that the dye adsorption efficiency gradually decreased with increasing dye concentration until the adsorption equilibrium condition was reached. The MG removal efficiency on SB cellulose decreased from 99 to 96%. The equilibrium condition was reached within 1 h. Figure 4 showed the plot between initial dye concentration and the percentage of MG dye removed.

Fig. 4 Effect of initial dye concentration on % removal

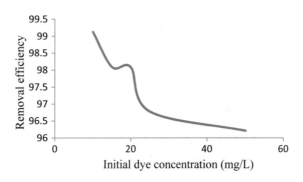

Fig. 5 Effect of system
temperature on % removal

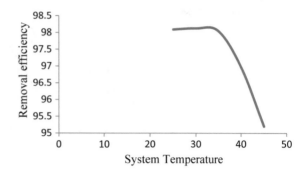

3.6 Effect of Temperature

In general, the dye removal efficiency increased with the increasing system temperature. In our project work, it was found that the removal efficiency decreased from 98.14 to 95% with the increasing temperature from 25 to 45 °C. Figure 5 showed that maximum MG dye adsorption on SB cellulose was 98.14% at 30 °C. It is well known that the motility of dye molecules may be enhanced by raising the system temperature and it may also reduced active sites on adsorbent surface. At high temperature, uncontrolled molecular movement facilitated desorption process (Ghorai et al. 2014) which reduced the adsorption efficiency.

3.7 Adsorption Isotherm

Commonly used equilibrium adsorption isotherm models, Langmuir and Freundlich (Ghorai et al. 2014; Metin et al. 2013), were used to determine the adsorption data of MG on the SB cellulose. Empirical expression of the Langmuir model is given by following equation.

$$C_S/q = 1/(k \times Q_0) + C_s/Q_0$$

where q (mg/g) is represented as the amount of adsorbed dye per unit mass of adsorbent and C_s (mg/L) is denoted as the unabsorbed dye concentration in solution at equilibrium condition. Q_0 is indicated as the maximum monolayer capacity at equilibrium condition and k presented as Langmuir constant (L/mg).

Mathematical expression of the Freundlich isotherm is given by following equation.

$$\log q = \log k_F + (1/n) \log C_s$$

Fig. 6 Adsorption isotherm by using Freundlich model

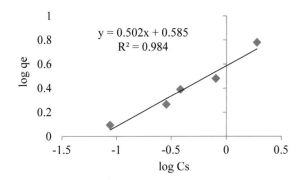

Table 1 Calculated parameters of Langmuir Isotherm	Temp (°C)	k (L/g)	Q_0 (mg/g)	R^2
	30	1.133173	8.161	0.9082

Table 2 Calculated parameters of Freundlich Isotherm	Temp (°C)	K_F (L/mg)	n	R^2
	30	3.851	1.989	0.984

where k_F and n are Freundlich constants. The value of K_F (mg/g(L/mg)$^{1/n}$) is indicated the adsorption capacity of the adsorbent. Values of $n > 1$ is indicated the condition of adsorption favorability (Ncibi et al. 2007).

Figure 6 showed Freundlich isotherm model for MG adsorption by SB cellulose. Calculated parameters for Langmuir isotherms and Freundlich isotherm models are shown in Tables 1 and 2. The adsorption of MG dye on SB cellulose fitted the Freundlich model very well with the correlation coefficient $R^2 = 0.984$, better than the Langmuir Isotherm with $R^2 = 0.9082$. Malachite green dye adsorption system process using SB Cellulose well fitted by Freundlich isotherm due to heterogeneous distribution of surface active side on SB cellulose (Vimonses et al. 2009). A value of 1.989 obtained for n indicates the condition of favorable adsorption process (Hameed 2009).

3.8 Recyclability Study

In evaluation of economy in adsorbent application, recyclability and reuse are important factors. Experiment on the recyclability of the SB cellulose adsorbent was done at 30 °C. 8 g/L of SB cellulose adsorbents were added into 50 mL of 20 mg/L dye solutions for 60 min and then taken out from the solutions. After adsorption, the MG dye adsorbed SB cellulose was added into 0.1 M NaOH or 0.1 M HCl solution and was stirred for 60 min at 30 °C. Then, The SB cellulose was washed with deionized water two times to remove the residual NaOH or HCl solution. Then, it was dried by vacuum dryer at 80 °C and reused for adsorption. By this process, generated adsorbents were reused for another adsorption process in the subsequent cycles. The adsorption efficiency of the SB cellulose for MG removal was still over 80% after the fourth cycle.

4 Conclusion

Extraction of cellulose from sugarcane bagasse was performed successfully. Malachite green dye removal efficiency on cellulose obtained from sugarcane bagasse was also investigated by adsorption study. It was observed that the removal efficiency depends on the parameters, as in: adsorbate concentration, adsorbent mass, contact time, pH, temperature, and rotational speed. The maximum MG dye removal efficiency of SB cellulose was observed as 98% for 20 mg/L dye solution using 8 g/L of adsorbent within 60 min at 30 °C and neutral pH. Modeling of experimental data was evaluated through two isotherm models: Langmuir and Freundlich isotherm model. It was found that Freundlich isotherm model fitted well for this adsorption process on SB cellulose than Langmuir isotherm model because of heterogeneous surface active side distribution on adsorbent surface. FTIR spectra of SB cellulose indicated the presence of hydroxyl groups, lignin structure, and hemicelluloses structure on the surface of adsorbents. From the results, it can be concluded that extracted SB cellulose was a stable and efficient adsorbent for removal of MG dye from their aqueous solution.

Acknowledgement The authors would like to thank the Dalmia Holding Groups for the financial support and also thankful to the Department of Chemical Engineering, Jadavpur University for great facilities and support.

References

Attallah, M. F., Ahmed, I. M., & Hamed, M. M. (2013). Treatment of industrial wastewater containing Congo Red and Naphthol Green B using low-cost adsorbent. *Environmental Science and Pollution Research, 20,* 1106–1116.

Garside, P., & Wyeth, P. (2003). Identification of cellulosic fibres by FTIR spectroscopy: Thread and single fibre analysis by attenuated total reflectance. *Studies in Conservation, 48,* 269–275.

Ghorai, S., Sarkar, A., Raoufi, M., Panda, A. B., Schonherr, H., & Pal, S. (2014). Enhanced removal of methylene blue and methyl violet dyes from aqueous solution using a nanocomposite of hydrolyzed polyacrylamide grafted xanthan gum and incorporated nanosilica. *ACS Applied Materials & Interfaces, 6,* 4766–4777.

Hameed, B. H. (2009). Spent tea leaves: A new non-conventional and low-cost adsorbent for removal of basic dye from aqueous solutions. *Journal of Hazardous Materials, 161,* 753–759.

Hu, Y. Q., Guo, T., Ye, X. S., Li, Q., Guo, M., Liu, H. N., et al. (2013). Dye Adsorption by resins: Effect of ionic strength on hydrophobic and electrostatic interactions. *Chemical Engineering Journal, 228,* 392–397.

Jacobsen, S. E., & Wyman, C. E. (2002). Xylose monomer and oligomer yields for uncatalyzed hydrolysis of sugarcane bagasse hemicellulose at varying solids concentration. *Industrial and Engineering Chemistry Research, 41,* 1454–1461.

Khalil, H. P. A., Ismail, H., Rozman, H. D., & Ahmad, M. N. (2001). The effect of acetylation on interfacial shear strength between plant fiber and various matrices. *European Polymer Journal, 37,* 1037–1045.

Kumar, A., Singh Negi, Y., Choudhary, Y., & Bhardwaj, N. K. (2014). characterization of cellulose nanocrystals produced by acid-hydrolysis from sugarcane bagasse as agro-waste. *Journal of Materials Physics and Chemistry, 2,* 1–8.

Ma, W., Song, X. Y., Pan, Y. Q., Cheng, Z. H., Xin, G., Wang, B. D., et al. (2012). Adsorption behavior of crystal violet onto opal and reuse feasibility of opal-dye sludge for binding heavy metals from aqueous solutions. *Chemical Engineering Journal, 193–194,* 381–390.

Mandal, A., & Chakrabarty, D. (2011). Isolation of nanocellulose from waste sugarcane bagasse (SCB) and its characterization. *Carbohydrate Polymers, 86,* 1291–1299.

Mansouri, J., Harrisson, S., & Chen, V. (2010). Strategies for controlling biofouling in membrane filtration systems: Challenges and opportunities. *Journal of Materials Chemistry, 20,* 4567–4586.

Metin, A. U., Ciftci, H., & Alver, E. (2013). Efficient removal of acidic dye using low-cost biocomposite beads. *Industrial and Engineering Chemistry Research, 52,* 10569–10581.

Mittal, A., Mittal, J., Malviya, A., Kaur, D., & Gupta, V. K. (2010). Adsorption of hazardous dye crystal violet from wastewater by waste materials. *Journal of Colloid and Interface Science, 343,* 463–473.

Nada, A.-A. M. A., & Hassan, M. L. (2003). Phosphorylated cation-exchangers from cotton stalks and their constituents. *Journal of Applied Polymer Science, 89*(11), 2950–2956.

Ncibi, M. C., Mahjoub, B., & Seffen, M. (2007). Kinetic and equilibrium studies of methylene blue biosorption by *Posidonia oceanica* (L.) fibres. *Journal of Hazardous Materials, 139,* 280–285.

Ren, Y., Abbood, H. A., He, F., Peng, H., & Huang, K. (2013). Magnetic EDTA-modified chitosan/SiO_2/Fe_3O_4 adsorbent: Preparation, characterization, and application in heavy metal adsorption. *Chemical Engineering Journal, 226,* 300–3011.

Sain, M., & Panthapulakkal, S. (2006). Bioprocess preparation of wheat straw fibers and their characterization. *Industrial Crops and Products, 23,* 1–8.

Salas, C., Nypelö, T., Abreu, C. R., Carrillo, C., & Rojas, O. J. (2014). Nanocellulose properties and applications in colloids and interfaces. *Current Opinion in Colloids & Interface Science, 19,* 383–396.

Saliby, I. E., Erdei, L., Kim, J. H., & Shon, H. K. (2013). Adsorption and photocatalytic degradation of methylene blue over hydrogen–titanate nano fibres produced by a peroxide method. *Water Research, 47,* 4115–4125.

Silva Filho, E. C., Lima, L. C., Silva, F. C., Sousa, K. S., Fonseca, M. G., & Santana, S. A. (2013). Immobilization of ethylene sulfide in aminated cellulose for removal of the divalent cations. *Carbohydrate Polymers, 92,* 1203–1210.

Singh, K. P., Gupta, S., Singh, A. K., & Sinha, S. (2011). Optimizing adsorption of crystal violet dye from water by magnetic nanocomposite using response surface modeling approach. *Journal of Hazardous Materials, 186,* 1462–1473.

Song, Y., Ding, S., Chen, S., Xu, H., Mei, Y., & Ren, J. (2015). Removal of malachite green in aqueous solution by adsorption on sawdust. *Korean Journal of Chemical Engineering, 32,* 2443–2448.

Šturcová, A., Davies, G. R., & Eichhorn, S. J. (2005). Elastic modulus and stress-transfer properties of tunicate cellulosewhiskers. *Biomacromolecules, 6,* 1055–1061.

Sun, J. X., Sun, X. F., Zhao, H., & Sun, R. C. (2004). Isolation and characterization of cellulose from sugarcane bagasse. *Polymer Degradation and Stability, 84,* 331–339.

Sun, X. F., Xu, F., Sun, R. C., Fowler, P., & Baird, M. S. (2005). Characteristics of degraded cellulose obtained from steam-exploded wheat straw. *Carbohydrate Research, 340,* 97–106.

Tashiro, K., & Kobayashi, M. (1991). Theoretical evaluation of three-dimensional elastic constants of native and regenerated celluloses: Role of hydrogen bonds. *Polymer, 32,* 1516–1526.

Troedec, M., Sedan, D., Peyratout, C., Bonnet, J. P., Smith, A., Guinebretiere, R., Gloaguen V., et al. (2008). Influence of various chemical treatments on the composition and structure of hemp fibers. *Composites Part A-Applied Science and Manufacturing, 39,* 514–522.

Vimonses, V., Lei, S., Jin, B., Chow, C. W. K., & Saint, C. (2009). Kinetic study and equilibrium isotherm analysis of Congo Red adsorption by clay materils. *Chemical Engineering Journal, 148,* 354–364.

Wang, W. M., Cai, Z. S., Yu, J. Y., & Xia, Z. P. (2009). Changes in composition, structure, and properties of jute fibers after chemical treatments. *Fibers and Polymers, 10,* 776–780.

Wulandari, W. T., Rochliadi, A., & Arcana, I. M. (2016). Nanocellulose prepared by acid hydrolysis of isolated cellulose from sugarcane bagasse. *Materials Science and Engineering, 107,* 12–45.

Yagub, M. T., Sen, T. K., Afroze, S., & Ang, H. M. (2014). Dye and its removal from aqueous solution by adsorption: A review. *Advances in Colloid and Interface Science, 209,* 172–184.

Zhou, Y., Zhang, M., Wang, X., Huang, Q., Min, Y., Ma, T., et al. (2014). Removal of crystal violet by a novel cellulose-based adsorbent: Comparison with native cellulose. *Industrial and Engineering Chemistry Research, 53,* 5498–5506.

Status of Sewage Treatment in Bihar and Needs for Improvement

Prabhansu, Krishna Kant Dwivedi, Malay Kr Karmakar
and Pradip Kr Chatterjee

Abstract With the ever increasing growth in population across the nation, there is a need for proper disposal of solid waste of households. A comprehensive literature survey was conducted to study the status of sewage treatment plants (STP) in different states of India. More emphasis was laid on its status in different districts of Bihar. Existing plants are incapable of handling huge inflow of wastes through sewage pipe lines. If properly developed, STP could not only remove the problem of waste management but it will also provide financial gains. Biogas produced in most of these plants from the digesters is flared off and are not being stored, if done will certainly reduce the cost of cooking fuel. Only primary and secondary treatment of the plants is being done in most of the sewage treatment plants. If tertiary treatment is also provided, this might help in solving the problem of safe drinking water. Conventionally, these left over sludge from the digester still contains a lot of carbonaceous content which can be harnessed by converting them in the form of powders and be utilized in different gasifiers and pyrolyser to produce clean combustible gas. Several districts do not even have a single sewage treatment plant and people are forced to use pit latrines. It was found that districts like Buxar, Muzaffarpur and Bhagalpur are having very old STP which date back to 1885 AD and are insufficient to handle the load of rapidly increasing population. Government initiatives are required not only to properly manage it but also to utilize it in a more effective manner.

Keywords Waste management · Sewage treatment · Sludge · Gasification · Pyrolysis

Prabhansu (✉)
Mechanical Engineering Department, Muzaffarpur Institute of
Technology, Muzaffarpur, Bihar 842003, India

K. K. Dwivedi
Mechanical Engineering Department, National Institute of
Technology Durgapur, Durgapur, West Bengal 713209, India

M. K. Karmakar · P. K. Chatterjee
Energy Research and Technology, CSIR-Central Mechanical Engineering Research
Institute, Durgapur, West Bengal 713209, India

© Springer Nature Singapore Pte Ltd. 2020
S. K. Ghosh et al. (eds.), *Recent Trends in Waste Water Treatment and Water
Resource Management*, https://doi.org/10.1007/978-981-15-0706-9_3

1 Introduction

The main purpose of sewage treatment plants construction is to transform the household sewage into an easily manageable waste and the secondary purpose is to retrieve and to re-use the water after treatment of sewage. Untreated sewage contains water; nutrients that are mainly nitrogen and phosphorus; solids, pathogens that includes bacteria, viruses and protozoa; helminthes such as metals, different gases and Oil Street dust parking lots and buildings with toxic chemicals and chlorinated organic.[1] Untreated sewage that comes from rivers and oceans creates threat for human beings and environmental issues.[2] They cause acute illnesses including dysentery, diarrhoea, typhoid, cholera, and hepatitis A in human while direct sewage discharge in river and oceans leads to eutrophication and algal blooms reducing the oxygen content in water.[3] The sewage treatment plant should produce water that is safe for human consumption, to get water that can be used for the consumer with reasonable cost.

2 Current Status of STPs Across the Globe and in India

Latest technologies for water treatment are being used by some of the advanced nations of the world. The ozone technology is being used by United States, whereas AAO and oxidation ditch methods are commonly used in China.[4] In Germany, Federal Water Act (WHG) restricts wastewater discharge into river bodies which leads to that European country which has the highest waste water reprocessing and recycling rate. In Germany, more than 96% of the sewage waste water from households and public facilities are discharged into nearby STP for treatment.[5] Treated waste-water is widely used for irrigation, particularly in the Gulf countries. Currently, an average of 700,000 cubic meters per day is being used to meet irrigation and industrial demand in the countries such as Kuwait, Oman, UAE and Saudi Arabia.[6]

According to Pollution Control Board of India, 816 STPs are located in different States/UTs in the country, out of which only 522 STPs are operational. Table 1 describes the current status of STPs in different states, where inference can be drawn that Maharashtra and Gujarat have the maximum capacity of municipal STPs. It can also be correlated that the developed states are having more number of STPs, so STPs also act as growth parameter, which is quite obvious by the fact that wastes are more recycled and utilized in these states.

---

[1] www.cpcb.nic.in.

[2] www.indiaenvironmentportal.org.in/files/file/Status%20of%20STPs.

[3] http://cep.unep.org/publications-and-resources/marine-and-coastal-issues-links/wastewater-sewage-and-sanitation.

[4] https://www.sciencedirect.com/science/article/pii/S0160412016301040.

[5] http://www.bmub.bund.de/en/topics/water-waste-soil/water-management/wastewater/.

[6] https://link.springer.com/chapter/10.1007/978-1-4615-3282-8_46.

Table 1 State wise status of STPs across different states in India

S. No.	State/UT	Capacity of municipal STPs	No. of municipal STPs	Operational capacity (MLD)	No. of STPs operational	Non-operational capacity (MLD)	No. of STPs non-operational	Under construction capacity (MLD)	No. of STPs under constructions	Proposed capacity (MLD)	No. of STPs proposed
1	Andhra Pradesh	247.27	12	156.27	9	–	–	91	3	–	–
2	Arunachal Pradesh	–	–	–	–	–	–	–	–	–	–
3	Andaman & Nicobar Islands	–	–	–	–	–	–	–	–	–	–
4	Assam	0.21	1	0.21	1	–	–	–	–	–	–
5	Bihar	124.55	6	99.55	5	25	1	–	–	–	–
6	Chandigarh	314.5	5	314.5	5	–	–	–	–	–	–
7	Chhattisgarh	–	–	–	–	–	–	–	–	–	–
8	Delhi	2693.7	35	2671.2	34	22.5	1	–	–	–	–
9	Daman & Diu Dadra Nagar Haveli	–	–	–	–	–	–	–	–	–	–
10	Goa	74.58	7	34.5	4	–	–	40.08	3	–	–
11	Gujarat	3062.92	51	2111.64	32	498	4	359.5	8	93.78	7
12	Haryana	852.7	41	805	38	2.7	2	45	1	–	–
13	Himachal Pradesh	114.72	66	79.51	36	35.21	30	–	–	–	–
14	Jammu & Kashmir	264.74	19	145.74	15	2	1	117	3	–	–

(continued)

Table 1 (continued)

S. No.	State/UT	Capacity of municipal STPs	No. of municipal STPs	Operational capacity (MLD)	No. of STPs operational	Non-operational capacity (MLD)	No. of STPs non-operational	Under construction capacity (MLD)	No. of STPs under constructions	Proposed capacity (MLD)	No. of STPs proposed
15	Jharkhand	117.24	15	117.24	15	–	–	–	–	–	–
16	Karnataka	1304.16	57	1112.05	44	–	–	192.11	13	–	–
17	Kerala	152.97	10	112.87	6	3	1	37.1	3	–	–
18	Lakshadweep	–	–	–	–	–	–	–	–	–	–
19	Maharashtra	5160.36	76	4683.9	60	344.5	10	131.96	6	–	–
20	Madhya Pradesh	482.23	17	475.48	14	6.75	3	–	–	–	–
21	Manipur	–	–	–	–	–	–	–	–	–	–
22	Meghalaya	1	1	–	–	1	1	–	–	–	–
23	Mizoram	10	1	–	–	–	–	10	1	–	–
24	Nagaland	–	–	–	–	–	–	–	–	–	–
25	Odisha	385.54	13	158.04	7	–	–	227.5	6	–	–
26	Puducherry	68.5	6	17.5	3	–	–	51	3	–	–
27	Punjab	1245.45	86	921.45	38	15.2	4	276.7	31	32.1	13
28	Rajasthan	865.92	63	384.5	16	–	–	149.3	11	332.12	36
29	Sikkim	31.88	11	8	1	5	1	18.88	9	–	–
30	Tamil Nadu	1799.72	73	1140.83	33	5.17	1	521.08	28	132.64	11
31	Telangana	685.8	18	634.8	17	–	–	51	1	–	–
32	Tripura	0.05	1	0.045	1	–	–	–	–	–	–

(continued)

Table 1 (continued)

S. No.	State/UT	Capacity of munic- ipal STPs	No. of munici- pal STPs	Operational capacity (MLD)	No. of STPs opera- tional	Non- operational capacity (MLD)	No. of STPs non- operational	Under construc- tion capacity (MLD)	No. of STPs under construc- tions	Proposed capacity (MLD)	No. of STPs proposed
33	Uttar Pradesh	2646.84	73	2372.25	62	89.59	7	170	3	15	1
34	Uttarakhand	152.9	24	90.75	10	–	–	39.15	12	23	2
35	West Bengal	416.9	28	235.36	16	181.54	12	–	–	–	–
	Total	23277.36	816	18883.2	522	1237.16	79	2528.36	145	628.64	70

http://www.bmub.bund.de/en/topics/water-waste-soil/water-management/wastewater/sewage-treatment-plants/

3   Current Status of STPs in Bihar

Bihar is one of the states that are facing worst condition as evident from the Table 1. In the entire state, only 06 STPs are installed. These STPs are very old and dates back to British rule in India. Prominent towns & cities have been considered for studying the status of STP, out of which Bhagalpur, Begusarai, Hajipur, Buxar and Patna are the prominent ones.

3.1   Sewage Project, Bhagalpur

Water supply system is very poor and identified as a major problem for Bhagalpur town. The current water treatment plant (WTP) in Barari, commissioned in 1885 has design capacity of 17.3 million liters per day (MLD) but it operates only at 12 MLD due to deterioration of the plants and machineries. Underground sewerage network is completely absent in Bhagalpur town. About 70% of the households are using septic tanks, but still around 20% are forced to use pit latrines. The remaining 10% have no other choice but to do open defecation. Wastewater management system in Bhagalpur was built under Ganga Action Plan for reducing the pollution level of the holy river. The total quantity of sewage water that reaches the sewage treatment plant is about 8.25 MLD. No reuse of the treated effluent is performed. Government is planning for the installation of sewage treatment of 47 MLD, funded through the World Bank.[7]

3.2   Sewage Project, Begusarai[8]

Sewerage system does not exist in this town. People are forced to perform open defecation. Households that are having sufficient area and in some of the newly developed colonies, septic tanks are used for waste water disposal and effluents arising from these septic tanks are discharged in open drains which further gets collected in local ponds. A sewage utilization plant with 17 MLD capacity of waste has proposed of 105 km length. Begusarai museum and the Gayatri shakti peeth are the sights for the two pumping stations. The proposed treatment process shall be Activated Sludge process which will be based on Sequential Batch Reactor technology. The STP site will be south east of the town near Wajidpur and the after treatment, effluent shall be allowed to get discharged in river Ganga.

---

[7]http://igemportal.org/Resim/Wastewater%20Treatment%20Technologies_%20A%20general%20rewiev.pdf.

[8]http://chinawaterrisk.org/resources/analysis-reviews/8-facts-on-china-wastewater/.

3.3   Sewage Project, Buxar and Hajipur[9]

Bihar government launched different Sewerage projects such as Buxar, Hajipur and Begusarai near Ganga river for improvement this project and financial support. The main objective of this project is rejuvenation of Ganga water quality by preventing from municipal sewage into the river.

The authority board of Ganga river has decided that municipal sewage and industrial waste would be enters Ganga river after 2020 for the project clean Ganga mission. Total Rs. 15,000 crore investment is needed for coming 10 years for the necessary treatment and sewerage infrastructure.

3.4   Sewage Treatment Plant, Patna

Next is the capital of Bihar state, Patna which is the second largest city in eastern India, after Kolkata. Sewerage system was first introduced in Patna town during the year 1936-39 under which a STP of 4.5 MLD was constructed in the year 1937 at Saidpur. At present, the Municipal limits of PMC from Patna Urban Agglomeration Area. The PMC cover an area of 100 km^2 with the present population of 16.83 lakhs as per 2011 census. Patna city covering an area of 100 km^2 is subdivided into 6 Sewerage Zones: Digha, Beur, Saidpur, Kankarbagh, Pahari and Karmali Chak. Saidpur zone covers the central part of Patna City. In this area, the population is very dense therefore, the sewage generation is high. The old Saidpur Zone of Patna as per the existing sewerage scheme is now divided in two zones namely Zone-III & IV A (N). This zone have a STP within its boundary at Saidpur STP (45 MLD plant) site.[10] At present, Saidpur zone has an existing ASP based STP capacity of 45 MLD which is catering a load of 33 MLD with primary and secondary level of treatment which is proposed to be augmented for treatment of 60 MLD. For proper maintenance of the sewerage system, provision has been made for one jetting cum suction machine, submersible dredger pump, safety equipment, and sewer cleaning machine of power bucket type which will help to collect the sludge and will disposed on identifies PMC waste disposal site regularly by packed tractor trolley, mounted tanker and other environmental friendly collection and disposal sources. At present, the technology used is Conventional Activated Sludge Process (ASP). The merits of this method are that the land requirement is less compared to others. Also better control is possible with reduced flies and odour nuisance. At the same time it involves high capital cost, high power requirement and skilled labour. With the present population of 3.64 lakhs which generates sewage discharge of 43.69 MLD, open drains are acting sewers during the dry weather flow.

---

[9]http://chinawaterrisk.org/resources/analysis-reviews/8-facts-on-china-wastewater/.

[10]http://biomasspower.gov.in/document/download-lef-tside/Biomass%20gasification.

3.4.1   Problems Identified

As per the visit of the Saidpur Sewage Treatment Plant, some problems with the typical plant of Bihar are identified as under:

- The methane gas produced as a result of decomposition of the sludge is completely flared off into the atmosphere. This can be brought up into use by collecting it in a gas dome but there was no mechanism for that.
- Since the plant is under construction for the expansion of capacity, area which has to be used for sun drying is now being occupied by the government for infrastructure development. Now the digested sludge cannot be sun dried and is now left over in the tank itself which produces risks of blockage of channels.
- The water even after treatment could not meet the requisite standards. This is due to workers' inefficiency and lack of proper reviewing policy.
- Lack of flow of sewage in the absence of sufficient sewer length in the drainage system.
- Power shortage is one of the factors which is being faced by the existing STPs.
- Shortage of manpower both skilled and unskilled type has led to mismanagement.
- Government is not providing enough funds for the basic requirements i.e., repair of machines, etc.

3.4.2   Recommended STP Technology

This SBR based project effluent can be beneficial for utilization for non-domestic purposes such as house flushing, gardening and floor cleaning. The effluents from sewage can be dried and the separated from sewage. After drying, pretreatment process can be applied and that processed sewage sludge may be used as biomass for the gasification plants. Gasification of biomass is a well-known technology and for this different gasifier is available with different designs and capacities according to the requirements. A biomass gasifier is similar as other renewable sources like solar energy and wind energy means it can produce power when required.[11] Whereas large thermal power plants and solar and wind based units are very location specific, Sewage sludge gasifier based systems can be set up at almost any place. Sewage sludge is a CO_2 neutral fuel and, therefore, unlike fossil fuels such as diesel does not contribute to net CO_2 emissions, which makes biomass based power generation systems an attractive option in mitigating the adverse effects of climate change.[12]

---

[11] https://energypedia.info/wiki/Biomass_Gasification_(Small-scale).

[12] https://en.wikipedia.org/wiki/Gasification.

4 Conclusions

Bihar state has very few STP and the existing ones are in very poor condition. It can be concluded that the Saidpur STP is working well but still needs some major modifications. The plan of expansion of capacity of the plant from 45 MLD to 60 MLD has affected the regular working to a large extent. There are definitely some implementation shortages which needs an eye by the officials. The overall operations at various stages are its fine screen channels, PSTs or secondary sedimentation tanks needs more labor concern as the final treated water was not up to the set standards. The biomass gasification technology of sewage sludge is theoretically an interesting option for rural development. Dangerous threats exist to the environment and health due to carcinogenic waste involved in the sewage waste. Therefore, at present the application of the gasifier technology for small-scale electricity production in developing countries seems to be justifiable. Each new plant would be a unique tailor-made facility.

Effluent Water Treatment: A Potential Way Out Towards Conservation of Fresh Water in India

S. Bej, A. Mondal and P. Banerjee

Abstract In modern age, rapid industrialization catalyses the steady magnification of the adverse effects. Released wastewater from various industries is of deepest concern which is highly responsible for aquatic pollution especially in India. An estimated amount of ~61,754 million litres per day (MLD) effluent water is produced in foremost metropolitans in India; however, the sewage management facility is only of ~22,963 MLD (approximately 37% of industrial effluents), and the remaining is disposed of untreated (according to the 2015 report of the Central Pollution Control Board 2015). The unscientific way of disposal causes severe diseases, responsible for environmental and ecological imbalance. This aquatic pollution is heading towards moderate to severe water shortages, brought on by the simultaneous effects of agricultural growth, industrialization and urbanization. In the forthcoming decades, the crisis of freshwater may lead to a major societal problem and political instability, if not the new option is found to supply the clean water. The development of cost-effective and eco-friendly wastewater management technologies and recycling could be the best way to solve the problem of potable water scarcity. Several conventional treatment technologies, i.e. adsorption, chemical coagulation, activated sludge management, membrane filtration, etc., have been implemented to eliminate the pollutants from effluent waters. In addition, it is straightforward, with good effectiveness and ability for degrading contaminants. This paper emphasizes the recent advancement and simultaneous use of wastewater treatment technologies in India to remove pollutants from wastewater like halogenated hydrocarbons, heavy metals, dyes and pigments, pesticides, herbicides, etc., which correspond to the main pollutants in wastewater. Movements like Swachh Bharat Mission (SBM), which has become the

S. Bej · A. Mondal · P. Banerjee (✉)
Surface Engineering & Tribology Group, CSIR-Central Mechanical Engineering
Research Institute (CMERI), Mahatma Gandhi Avenue, Durgapur, West Bengal, India
e-mail: pr_banerjee@cmeri.res.in

S. Bej · P. Banerjee
Academy of Scientific and Innovative Research at CSIR-Central Mechanical Engineering
Research Institute (CMERI), Mahatma Gandhi Avenue, Durgapur, West Bengal, India

A. Mondal
Department of Chemistry, National Institute of Technology, Mahatma Gandhi
Avenue, Durgapur, West Bengal, India

© Springer Nature Singapore Pte Ltd. 2020
S. K. Ghosh et al. (eds.), *Recent Trends in Waste Water Treatment and Water
Resource Management*, https://doi.org/10.1007/978-981-15-0706-9_4

largest movement towards cleanliness, highly emphasizes the treatment along with recycling of these effluent waters throughout India.

Keywords Effluent water · Water pollutants · Heavy metals · Toxicity · Natural adsorbents · International society of waste management · Air and water

1 Introduction

Rapid industrialization and urbanization have been adversely affecting the global environment in addition to that the quality as well as quantity of the potable water resources for the past few decades. Pollution by inappropriate industrial wastes management is one of the chief environmental problems in India. The basic technical knowledge required to cope up with pollution is already known to us, but unfortunately, less concern about the fact leads us towards self-destruction.

Access of clean water is immensely indispensible for growth and development, as water is a lifeline for all forms of life on this earth, and it is basically acquired from two principal usual sources, natural surface water like rivers, freshwater lakes, streams, etc., and groundwater such as borehole water and well water. So, contamination of these water bodies creates a real day crisis of freshwater. Generally, the industrial development occurring along the banks of rivers exerts pressure on the water resources causing the shortage and contamination of surrounding water bodies. Moreover, disposal of untreated wastewater directly into the water bodies causes the surface and groundwater pollution. Normally, wastewater consists of ~99.90% water and ~0.10% suspended or dissolved solids mainly bio-organic in nature (human waste, vegetable matter, paper, etc.), various toxic metals and pathogens which consume available oxygen from water bodies resulting in the increase of BOD and COD level. The contamination of water bodies by toxic metals is a severe problem for all living organisms. The metals get bio-accumulated in the nature and consecutively get biomagnified next to the food chain. Though metals like Cu, Fe and Zn are vital micronutrients, but excess accumulation leads to several lethal effects (Kar et al. 2008; Nair et al. 2010). The heavy metals such as iron, cadmium, lead, zinc, chromium, vanadium, nickel, copper, platinum, silver, titanium, etc., are commonly released from various industrial sectors like tannery, textile, leather, pigment and dyes, paint, petroleum refining industries, etc. Thus, treating the effluent water in the industrial premises before discharging it into the environment is of extreme importance. Table 1 shows the maximum contaminant level (MCL) standards for some heavy metals established by USEPA (Tripathi and Ranjan 2015).

The amplifying demand of freshwater for various domestic purposes is met by groundwater, which is presently at high jeopardy due to continuous withdrawal. Treated wastewater can be a substitute to the groundwater for some domestic uses. Therefore, mitigation of water pollutants from industrial effluent becomes compulsory before disposing it directly into the environment. A series of approaches have

Table 1 The MCL values of the most hazardous heavy metals

Name of heavy metal	Toxicity	MCL (mg/L)
(i) Arsenic (As)	Skin manifestations, visceral cancers, vascular disease	5×10^{-2}
(ii) Cadmium (Cd)	Kidney damage, human carcinogen	1×10^{-2}
(iii) Chromium (Cr)	Headache, diarrhoea, nausea, vomiting, carcinogenic in nature	5×10^{-2}
(iv) Copper (Cu)	Liver damage, insomnia, Wilson Disease	25×10^{-2}
(v) Nickel (Ni)	Nausea, chronic asthma, coughing, carcinogen, Dermatitis	20×10^{-2}
(vii) Lead (Pb)	Damages the foetal brain, kidney diseases, circulatory system, nervous system	6×10^{-3}
(viii) Mercury (Hg)	Rheumatoid arthritis, kidney diseases, circulatory and nervous system	3×10^{-5}

been looked into such as coagulation, membrane technique, adsorption, foam flotation, reverse osmosis, catalytic degradation and other bio-techniques for the removal of contaminants from effluent water (Pontius 1990). Out of these processes, adsorption has come out to be one of the safest and cheapest alternative methods. Indian Government has already taken many steps to nurture the matter like 'Ganga rejuvenation' [a part of Swachh Bharat Mission (SBM)], according to the Environment Protection Act, 1986, as a planning, financing and monitoring for effective reduction of pollution and conservation of the river Ganga. This review article deeply concerns about the improper management of generated effluent water from industries in India and emphasizes the primary treatment systems available for treating wastewaters. To properly address the pollutants releasing from the industries, proper management of generated wastes and adoption of greener and cost-effective technology are being encouraged day by day.

1.1 A Historical Perspective

In the year 1900, scientist Louis Pasteur and other researchers showed that sewage-borne bacteria grow due to the contamination of human excrement and are responsible for severe infectious diseases (Henze and Harremoes 1983). Therefore, the contaminated water used to spread in large farms to be decayed by the action of micro-organisms. It was quickly investigated that the land converted to be 'sick' and unproductive. Later, attempts indulged direct disposal of effluents into the surface water which caused the aquatic pollution. But, it was soon realized that environment might not act as an unlimited dumping reservoir. Basic methods of effluent water treatment were first advanced in response to the adverse circumstances caused by the unscientific disposal of effluent water to the nature and public health. Furthermore, due to the urbanization, inadequate land was available for wastewater treatment and clearance. Moreover with population explosion, the quantity of effluent water produced per day increases rapidly and goes beyond the self-purification capacity (Spellman 1999; Rosen et al. 1998). Therefore, other treatment methods were developed and resulted an upliftment of the technologies from 1900 to the early 1970s that aims at-

(i) Removal of unwanted suspended materials from wastewater
(ii) Treatment of chemicals along with biodegradable organics (COD and BOD removal)
(iii) Inhibition of lethal pathogenic micro-organisms.

Major objective is to attain efficient and prevalent wastewater treatment technology to get better quality of water. Since 1990, effluent water treatment has been started by increasing scientific knowledge and various Government campaigns which are still continuing, but the emphasis has been shifted to the cause and removal of pollutants to improve the degree of treatment.

1.2 Distribution of Total Water on Earth's Surface

The circulation of total water on the Earth's surface is shown in Fig. 1, which states that only ~3% of water sources on the surface is fresh (among which, ~69% in glaciers, ~30% underground and ~1% in lakes, rivers and swamps); the remaining ~97% exists in the ocean. Hence, it is clear that only 1% of the water source on the surface of Earth is usable.

2 Category of Effluent Water

Effluent water is a side product of domestic, industrial, agricultural activities, surface runoff or any sewer infiltration (Tilley et al. 2016). The qualities of wastewater vary depending on the sources and the degree of contamination. It may be wastewater released from households, municipal wastewater (sewage) coming from societies or industrial wastewater from industrial sectors. Effluent water generally contains

Fig. 1 Distribution of earth's water

physical, chemical and biological pollutants. Basically, they are categorized into three broad classes: Black water, Grey water and yellow water (CPCB 2005).

2.1 Black Water

It is used to describe the effluents coming from the toilets, contaminated with various pathogens.

2.2 Grey Water

Grey water or Sullage is all kinds of effluent water generated from households or offices, except the wastewater from toilets. It contains lesser pathogens and therefore easier to treat and reuse for other purposes like sanitation, crop irrigation, etc. Reuse of grey water provides enormous benefits to reduce the demand of freshwater and minimizes the generation of wastewater.

2.3 Yellow Water

Water basically mixed with urine gathered from specific channels which are not polluted with either black water or grey water.

2.4 Sources of Effluent Water

Effluent waters originate from various sources like residences (household wastes), commercial sectors (schools, hospitals, offices, hotels, etc.) and industries. Apart from these, wastewaters are generated from non-pointed sources such as storm water, contaminated rainfall runoff from agricultural lands and entrance of the groundwater to the wastewater collection system. The resource of wastewater generally determines the qualities of the contaminated water, based on which the treatment procedures must be designed.

2.4.1 Domestic Sewage

This includes effluent water from homes and offices, typically contains contaminants such as vegetable, faecal matter, oil and grease, detergents, rags and sediment.

2.4.2   Industrial Wastewater

Industrial effluent water is one of the prime sources of water pollution, includes toxic chemicals, organic and radioactive wastes, large amount of sediments, high-temperature waste or acidic/caustic waste. During the last century, a large amount of industrial wastewater was disposed directly into the environment without any treatment. The main concern is that the local farmers unconsciously use this raw sewage to irrigate their lands, and the agricultural products are marketed locally. Therefore, the pollutants may indirectly enter the food chain causing serious problems to the eco-system and human's life. In Durgapur industrial belt, severe incidents took place due to drastic consequences of wastewater pollution (Javed et al. 2013). Some of them are mentioned below:

(i)   In 1973, thousands of fishes and frogs died, and ~240 hectares of cultivated lands were lost in the left bank irrigation canal at the Durgapur Barrage of the Damodar Valley Corporation (DVC).
(ii) In 1978, a huge number of cattle are died after drinking contaminated canal water.
(iii) In 1983, ~500 domestic animals died, and ~8000 quintals of processed crops were demolished. All the three catastrophes occurred due to the disposal of toxic chemicals into the Damodar canal water by the nearby small- and large-scale industries.

2.5   General Characteristics of Industrial Wastewater

A vivid idea of the chemical composition of wastewater is important since this helps to know the probable reactions with the organic and inorganic matters (Roila et al. 1994). Generally, the effluent water released from different industrial sectors can be separated into two broad categories: (i) inorganic industrial wastewater and (ii) organic industrial wastewater.

2.5.1   Inorganic Industrial Wastewater

This kind of wastewater is generated largely in the mining industries, nonmetallic minerals industries and metal surface processing industries (iron pickling, electroplating plants, etc.).

2.5.2   Organic Industrial Wastewater

Organic industrial effluents are released from several organo-chemical industries, which mainly use organic materials for reactions and contain organic contaminants such as phenol, organic sulphur, volatile organic matters, surfactants, etc. Table 2 describes the general parameters of effluent waters coming from different industrial sectors (Kaur et al. 2012).

The large amount of produced effluent water destroys the eco-system, mainly marine lives. Hence, it becomes a burning issue to purify the wastewater properly. Therefore, we should be keen to control the production of the wastewater, consumption of clean water in different sectors and cost of the treatment of wastewater (Fig. 2).

Table 2 Categories of industrial effluent water

Industrial sector	General contaminants
(i) Steel and Iron	High BOD and COD, oil and grease, high pH, acids, phenols, cyanide, heavy metals like Mn, As, Hg, Cd, Ti, V, Sb, Th, Pb
(ii) Leather and textiles	High BOD, solids, sulphates
(iii) Paper and pulp	High BOD and COD, solids, chlorinated organic compounds
(iv) Refineries and petrochemicals	Mineral oils, BOD, COD, phenols, chromium
(v) Chemicals	Cyanide, carbon-based chemicals, heavy metal ions, SS, COD
(vi) Non-ferrous metals	Fluorine, suspended solids
(vii) Microelectronics	Organic chemicals, COD
(viii) Mining	Suspended solids, metals, acids, salts

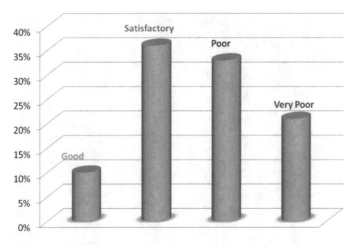

Fig. 2 Wastewater treatment plants in India under different grades

2.6   Present Status of Effluent Water Conundrum and Treatment in India

In India, climate alteration in the monsoon has huge implications on agriculture that makes India more vulnerable because >60% of countrymen are relying on cultivation for survival, and ~two-thirds of the nurtured terrestrial is fed by rain. Simultaneously, the government has launched ambitious programs like 'Smart City Mission', 'Swachh Bharat Mission', and the 'Make in India' campaign which needs to be reviewed from water and environmental aspects. Moreover in urban areas, freshwater resources are at high risk because of the increasing demand of freshwater. Apart from this, contaminants from various industrial sectors pollute the local water bodies making the freshwater endangered. Hence, a proficient solution is essential for viable water treatment. Therefore, reuse of purified effluent water would be of immense significance in achieving freshwater security and water conservation.

In India, a regular rainfall of ~1170 mm occurs corresponding to a yearly precipitation of ~4000 BCM (Billion Cubic Metre). However, based on the constraints like substantial variation in rainfall, evaporation throughout the year, etc., ~1123 BCM (690 from surface plus 433 (28%) from groundwater resources), gained from precipitation, is utilizable. Moreover, ~ 688 BCM of water is used for watering, which may upsurge to 1072 BCM in 2050. The average amount of ground water renewal is 433 BCM in a year, out of which ~212.5 BCM is utilized for watering, and ~18.1 BCM is spent in domestic and industrial usage (CGWB 2011). However, by 2025, the demand of water used in industrial and domestic purposes may rise to ~29.2 BCM (Fig. 3).

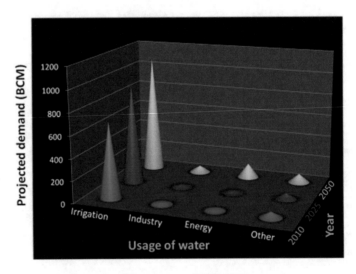

Fig. 3 Projected water demand at different sectors (CWC 2010)

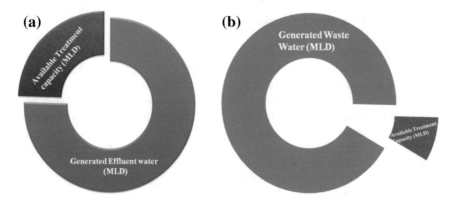

Fig. 4 Treatment capacity and sewage generation in class-I cities (**a**) and class-II townships (**b**) in India (CPCB 2009)

With rapid development of urbanization and domestic water usage, production of grey water is growing day by day. According to CPHEEO (The Central Public Health and Environmental Engineering Organization), ~70 to 80% of the water supplied for domestic usage transfers into wastewater. Survey reveals that wastewater production in class-I cities as well as class-II towns is ~98 lpcd (litres per capita per day). According CPCB (Central Pollution Control Board), the total generated wastewaters by 498 Class-I cities and 410 Class-II towns in India are ~35,558 and ~2696 MLD (millions of litres per day), respectively, while the installed sewage treatment capacity is just 11,553 and 233 MLD, respectively (Fig. 2). Five States in India, viz Delhi, Uttar Pradesh, Maharashtra, Gujarat and West Bengal are the foremost contributors of effluent waters (63%; CPCB 1999, 2007; CWC 2010). Further, according to WWAP and UNESCO (2006), it is anticipated that in 2050, ~48.2 BCM of wastewaters would be produced, which will further widen this gap (Bhardwaj 2005). Thus, entire study reveals that in near future, a double-edged problem will arise to deal with: reduced sources of pure water and increased effluent water production. Hence, there is an acute requirement for effectual water management (Fig. 4).

2.7 Wastewater Treatment Technologies

A host of new technologies for proper management of wastewater is being developed worldwide. However, new methodologies include natural procedures and are intended with feasibility. The techniques described below provide some basic idea on the miscellany, tractability and future utility.

2.7.1 Physicochemical Treatment Technologies

Physicochemical technique (Gupta et al. 2012) may be described as a treatment process in which biological and physical-chemical processes are mixed to get a hybridized treatment technique (Fig. 5).

The primary treatment includes decontamination and elimination of unwanted toxic elements depending upon some physical methods like screening, sedimentation, coagulation, flotation, etc. Secondary treatment deals with the biological process as aerobic, anaerobic, etc.

In the tertiary treatment, wastewater is converted into freshwater by using methods like distillation/evaporation/solvent extraction/advanced oxidation/precipitation/micro- and ultrafiltration, etc.

Adsorption is the most efficient and a universal technique used in the elimination of a wide variety of contaminants like colour, odour and various organic and inorganic impurities. The main advantage of this process lies in its low cost, straightforwardness and less sludge generation. Apart from this, reverse osmosis processes a widely used wastewater purification technique, but contains comparatively high cost for maintenance.

2.7.2 Bio-chemical Treatment Technology

Bio-chemical technology (Seow et al. 2016) including biofilm technology, aerobic granulation and bacterial fuel cell are a few useful methods for wastewater purification. Biofilm, a cluster of micro-organisms, is devoted to an exterior part. Dental plaque, for example, is one of the biofilms, comprises ~500 microbial species (MSU Center for Biofilm Engineering, 2008). The various species in biofilms, like, fungus, algae, yeasts, protozoa, etc., form groups by creating an 'extracellular polymeric substance (EPS)'. The cells develop three dimensional, hardy communities by making sugary molecular strands which excrete glue-like material to permit them to stick to an exterior of nearly any kind.

The biological wastewater treatment is often carried out by conventional treated sludge systems, which require bulky surface areas for management and biomass segregation units because of the deprived settling nature of the sludge.

Aerobic granules, having compactness, regularity, smoothness and good settling ability, can self-immobilize flocks and micro-organisms into spherical and strong compact structures.

Microbial fuel cell (MFC), a bio-chemical device, uses bacteria as a biocatalyst to transform chemical energy present in biological matter into electricity. MFC shows high energy conversion efficiency due to direct conversion of chemical energy within substrate to electricity.

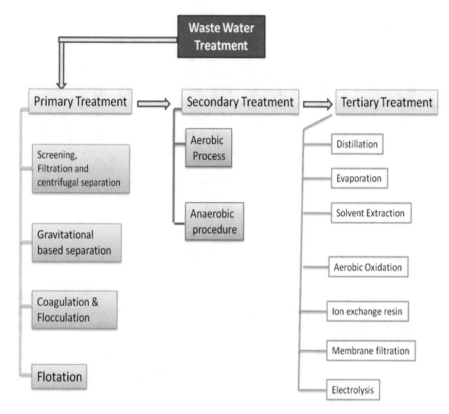

Fig. 5 Steps involved in chemical treatment methodologies are briefly demonstrated using a flow diagram

2.7.3 Bio Mediated Treatment Technology

Aquatic plant-based treatment systems (Dixit et al. 2011) using ponds or artificial wetlands are effective for purifying domestic and industrial drainage water. Biosorption is a proficient substitute of conventional techniques for elimination of noxious metals from effluents released from industries. Slower the flow and longer the retention time, greater will be the removal. Aquatic plants show remarkable absorption of metals (like N, P, S, Ca, Mg, Na, Fe, Mn, Zn, Cu, etc.) and nutrients. Removal of aquatic vegetation on timely basis is an important step; otherwise, died plants will decay releasing their absorbed nutrients to the water, thereby further degrading the water quality.

2.7.4 Membrane-Based Treatment Technology

Membrane-based separation technique is presently used as a replacement of conventional wastewater treatment processes and is the main market driver (Hamingerova et al. 2015). These technologies provide membrane separation based on selective separation through apertures of various sizes and contain four core membranes: microfiltration, ultrafiltration, nano-filtration (CNT-silicon nitride membranes, Ag particles impregnated membrane) and reverse osmosis (CPCB 2007; Noy et al. 2007). Micro and ultrafiltration are used for unwanted particle removal, whereas nano-filtration and reverse osmosis are used for softening or desalination. Handling of organic and inorganic contaminants typically requires modified sludge treatment by using a dipped membrane.

2.8 Proposed Wastewater Treatment Prototype

Because of the growing demand of potable water supply in the domestic sectors, there is an urgent requisite to prepare an inexpensive, gravitational-based domestic wastewater purification system (Fig. 6) that would convert contaminated water to

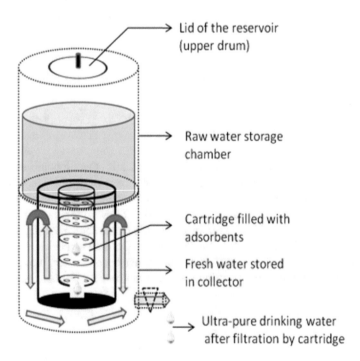

Fig. 6 Domestic wastewater treatment unit

potable water. The emphasis in this regard is to use naturally abundant adsorbent materials (e.g. sand, gravels, brick dust, etc.) for treatment purposes. The treated water will serve the required domestic purposes like gardening, washing, etc.

2.9 Conclusion

Clean potable water is indispensable for human activities. The demand of healthy drinking water is enhancing proportionally with increasing population. Hence, there is a pressing requirement for an effectual water management and wastewater recycling technology. In this milieu, nanotechnology is fast replacing traditional concepts, which depend on the improvement of newer multi-functional materials with unique properties and specific attraction towards certain contaminants. But, these approaches are convoluted, tedious and costly. Therefore, exploration is going on to develop inexpensive, but effective water treatment techniques. In a developing and heavily populated nation like India, large amount of effluent water is being produced from different sectors. This review article draws attention about concerning issues and technological options connected to the wastewater collection followed by treatment systems.

Acknowledgements The authors are thankful to the Director, CSIR-CMERI, Durgapur, for his immense support. SB is thankful to DST (vide ref. no: DST/TM/WTI/2K16/277(G) and project no: GAP-214312) for providing financial assistance. AM is thankful to UGC for her UGC sponsored NFSC fellowship.

References

Bhardwaj, R. M. (2005). Status of wastewater generation and treatment in India. *(IWG-Env) Joint Work Session on Water Statistics, Vienna*, 20–22.

CGWB. (2011). Ground water year book - India 2010–11. Central ground water board, ministry of water resources. Government of India. http://www.cgwb.gov.in/documents/Ground%20Water%20Year%20Book-2010-11.pdf.

CPCB. (1999). Control of urban population series. CUPS/ 1999-2000. Status of water supply and wastewater collection, treatment and disposal in class 1 cities.

CPCB. (2005). Performance status of common effluent treatment plants in India. Central Pollution Control Board, India.

CPCB. (2007). Control of urban pollution series: CUPS/68/2007. Evaluation of operation and maintenance of sewage treatment plants In India.

CPCB. (2009). Status of water supply, wastewater generation and treatment in class I cities and class II towns of India. Central Pollution Control Board, India. Series: CUPS/70/2009–10.

CWC. (2010). Water and related statistics. Water Planning and Project Wing, Central Water Commission, India.

Dixit, A., Dixit, S., & Goswami, C. S. (2011). Process and plants for wastewater remediation: A review. *Scientific Reviews and Chemical Communications, 1*(1), 71–77.

Gupta, V. K., Ali, I., Saleh, T. A., Nayak, A., & Agarwal, S. (2012). Chemical treatment technologies for waste-water recycling—An overview. *RSC Advances, 2,* 6380–6388.

Hamingerova, M., Borunsky, L., & Beckmann, M. (2015). Membrane technologies for water and wastewater treatment on the European and Indian market.

Henze, M., & Harremoes, P. (1983). Anaerobic treatment of wastewater, a literature review. *Water Science & Technology, 15,* 1–101.

Javed, A., Baranwal, A., Ishwarya, M. S., Ahmed, F., & Hidaytullah, Md. (2013). Bioremediation of industrial effluents of Asansol Durgapur industrial zone and its effect on DNA. *Global Jounal of Bio-science and Biotechnology, 2*(2), 215–226.

Kaur, R., Wani, S. P., Singh, A. K., & Lal, K. (2012). *Wastewater production, treatment and use in India* (pp. 1–4).

Kar, D., Sur, P., Mandal, S. K., Saha, T., & Kole, R. K. (2008). Assessment of heavy metal pollution in surface water. *International Journal of Environmental Science and Technology, 5*(1), 119–124.

Nair, I. V., Singh, K., Arumugam, M., Gangadhar, K., & Clarson, D. (2010). Trace metal quality of Meenachil River at Kottayam, Kerala (India) by principal component analysis. *World Applied Sciences Journal, 9*(10), 1100–1107.

Noy, A., Park, H. G., Fornasiero, F., Holt, J. K., Grigoropoulos, C. P., & Bakajina, O. (2007). Nanofluidics in carbon nanotubes. *Nanotoday, 2*(6), 21–29.

Pontius, F. W. (1990). *Water quality and treatment* (4th ed.). New York: McGraw-Hill Inc.

Rosen, M., Welander, T., Löfqvist, A., & Holmgren, J. (1998). Development of a new process for treatment of a pharmaceutical wastewater. *Water Science and Technology, 37,* 251–258.

Roila, T., Kortelainen, P., David, M. B., & Makinen, I. (1994). Acid-base characteristics of DOC in Finnish lakes. In: Senesi, N., Miano, T. M. (Eds.). *Humic substances in the global environment and implications for human health. Elsevier, Amsterdam,* 863–868.

Seow, T. W., Lim, C. K., Nor, M. H. M., Mubarak, M. F. M. M., Lam, C. Y., Yahya, A., et al. (2016). Review on wastewater treatment technologies. *International Journal of Applied Environmental Sciences, 11,* 111–126.

Spellman, F. R. (1999–2000). *Spellmann's standard handbook for wastewater operations* (Vols. 1, 2 and 3, pp. 60–80). Technomic Publishers: Lancaster, PA.

Tilley, E., Ulrich, L., Lüthi, C., Reymond, Ph., & Zurbrügg C., *Compendium of sanitation systems and technologies* (2nd Revised Edition). Swiss Federal Institute of Aquatic Science and Technology (Eawag), Duebendorf, Switzerland, p. 175. ISBN 978-3-906484-57-0. Archived from the original on 8 April 2016.

Tripathi, A., & Ranjan, M. R. (2015). Heavy metal removal from wastewater using low cost adsorbents. *Bioremediation & Biodegradation, 6,* 315.

Development of an Effective and Efficient Integrated Charcoal Filter Constructed Wetland System for Wastewater Treatment

V. V. D. N. G. Vidanage, A. K. Karunarathna, A. M. Y. W. Alahakoon and S. M. N. Jayawardene

Abstract Effective and efficient treatment of greywater from households and public places is an important and urgent issue to be solved in many developing countries. The higher capital investment of high-tech systems hinder the implementation, thus development of efficient yet affordable and compact wastewater treatment units for greywater treatment can solve the problem to a greater extend. This research developed a compact wastewater treatment system using commonly available construction and biomaterials. The proposed compact wastewater treatment system is consisted of three main components: up-flow anaerobic sludge blanket, charcoal granular media, and a subsurface flow constructed wetland system and is named Integrated Charcoal Filter Constructed Wetland (ICFiWet) system. In the vertical cylindrical reactor, the influent wastewater flows upward through an anaerobic chamber passes through the granular media to the subsurface flow wetland. The ICFiWet system was first tested at household scale (100 L capacity) fed with synthetic greywater. Water quality parameters such as pH, EC, TS, TSS, biological oxygen demand (BOD_5), nitrate N, and phosphate removals were estimated to evaluate the reaction kinetics and performances. The BOD removal was >66% and the system reduced the nitrate (1.77 ± 0.28) concentration by 15.3%. It was also revealed that the system decomposes the complex phosphorous substances to soluble P, leaving about 3 times increase of soluble P in effluent waters. The ICFiWet system was then up-scaled to 15.0 m^3 capacity to treat 5.0 m^3/day of flow. The ICFiWet system was installed in a larger public washroom to treat partially digested effluents. The results showed that the treated effluent achieved the water quality standard that comply with irrigation water quality. Thus, it was concluded that the ICFiWet system could effectively use for greywater treatment in household and public places while recovering the important nutrient phosphorus.

Keywords Greywater · Integrated · Up-flow · Charcoal · Constructed wetland · International society of waste management · Air and water

V. V. D. N. G. Vidanage · A. M. Y. W. Alahakoon · S. M. N. Jayawardene
Board of Study in Agricultural Engineering, Postgraduate Institute of Agriculture, University of Peradeniya, Peradeniya, Sri Lanka

A. K. Karunarathna (✉)
Department of Agricultural Engineering, Faculty of Agriculture, University of Peradeniya, Peradeniya, Sri Lanka

© Springer Nature Singapore Pte Ltd. 2020
S. K. Ghosh et al. (eds.), *Recent Trends in Waste Water Treatment and Water Resource Management*, https://doi.org/10.1007/978-981-15-0706-9_5

47

1 Introduction

Treatment of greywater from households and public places in developing countries is an urgent issue to be solved. Techniques of greywater treatment should fulfil five main criteria: hygienic safety, esthetic appearance, environment tolerance, technical and economic feasibility, and low cost, and the ability to easily maintain by unskilled workers (Nolde 2000; Gross et al. 2007). Implementation of greywater treatment systems has subsided due to economic and technical challenges (Ammari et al. 2014). Most of the available high-tech wastewater treatment systems are not affordable by the small wastewater generators, particularly households in developing countries because of high installation and maintenance costs (El-Khateeb and El-Gohary 2003). Therefore, greywater is generally disposed to the environment without any treatment as there is no any provision for its treatment (Katukiza et al. 2014).

Releasing wastewater to the environment without any treatment can cause various problems with regard to the environment and human health. Therefore, wastewater should be treated prior to discharge into environment. Generally, natural treatment systems are used for treating greywater as it is economically feasible and environmental friendly (Li et al. 2009; Boyjoo et al. 2013). The use of conventional low-tech wastewater treatment systems such as treatment wetlands and pond systems has gained limited attention due to space limitations in urban and semi-urban communities (Parkinson and Taylor 2003).

Wetlands and pond systems have gained limited attention due to space limitations in urban and semi-urban communities (Parkinson and Tayler 2003). Consequently, there is a demand for development of appropriate and innovative technologies for wastewater treatment and reuse for small communities (Bdour et al. 2009).

Constructed wetlands and granular filtration can be identified as extended natural greywater treatment systems which use natural media for biological degradation and filtration (Albalawneh et al. 2017). Among many small scale and low-cost alternatives (Kamal et al. 2008), constructed wetland system is one that highly preferred by small communities with limited funds. Those can be identified as the systems which have the ability to remove pollutants using interactions of plants and microorganisms (Li et al. 2004). Once the constructed wetlands are designed and established, maintenance costs are less. Also, they add an esthetic value to the environment (Lorion 2001). However, larger land footprint requirement and inability of conventional constructed wetlands to handle diverse range of wastewaters hinder the effective use of this affordable system.

Granular filtration achieves wastewater treatment through three main mechanisms: physical, chemical, and biological. Physical filtration, chemical transformation, and biological treatment due to microbial growth are occurred in granular filtration (Ciuk 2015). This technology is suitable for small communities in developing countries due to the low cost of construction, operation, and maintenance (Manz and Eng 2004). Cost effective and sustainable greywater treatment systems can be achieved by selecting locally available filter materials and using them in simple treatment methods (Nansubuga et al. 2015). Charcoal can be used as an effective filter material for greywater treatment (Dalahmeh et al. 2012).

Therefore, an attempt was made to integrate an up-flow anaerobic sludge blanket; a charcoal granular media and a constructed wetland system in one physical entity and developed a new concept: Integrated Charcoal Filter Constructed Wetland (ICFiWet) system. This study presents the design concepts, pilot-scale testing of the ICFiWet for synthetic greywater treatment, and preliminary results of the up-scaled design installed in a public place for secondary treatment of sewage wastewater.

2 Methodology

2.1 Pilot-Scale Design

The ICFiWet system consisted of three main components: an up-flow anaerobic sludge blanket, charcoal granular media, and a subsurface flow constructed wetland system. As shown in Fig. 1, three components were assembled vertically creating an updraft movement of wastewater through successive components.

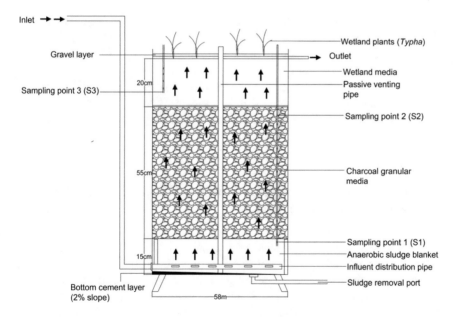

Fig. 1 Experimental setup of the pilot-scale reactor

2.1.1 Fabrication of the System

The pilot-scale reactor was established using an empty HDPE plastic barrel container having dimensions of 58 cm internal diameter and 92 cm height. A sludge removal port was fixed on the outside wall of the bottom, and 5-cm diameter perforated PVC pipes were fixed on the bottom to evenly distribute the influent. A perforated 4 mm thick plastic baffle was fixed 15 cm above the bottom, leaving the first chamber to be acting as an anaerobic reactor. A layer of softwood charcoal granules with particle sizes in between 0.5 and 3 cm was filled above the baffled up to 70 cm from bottom which acted as the charcoal granular bed. The constructed wetland media was placed above the charcoal bed, separated by a similar plastic baffle as above. Wetland media was prepared with top soil, sand, and charcoal mixture with 2:3:2 weight ratios. The wetland media was planted with *Typha angustifolia* plants which were collected from a microcosm previously grown in a nursery and washed with tap water to remove debris and dead parts.

Perforated stainless steel tubes (5 mm diameter) were vertically inserted into each component allowing water samples to be collected from different compartments.

2.1.2 Influent Greywater Preparation and Flow Rate

The influent for the system was a synthetic greywater solution prepared by mixing 500 g of kitchen/food waste, 8 g of laundry powder, 10 g of pulverized bar soap, 2.5 g of vegetable oils, and 100 ml of kitchen effluent in 100 L of tap water (Travis et al. 2010). A 100 L synthetic greywater solution was supplied to the system with a gravity head as shown in Fig. 1. With this arrangement, 18.3 L day^{-1} of influent passed through the system with a nominal residence time of 5.5 days. In order to avoid depletion of quality, the synthetic greywater solution was replaced every other day.

2.2 Upscaling of the Integrated Charcoal Filter Constructed Wetland (ICFiWet) System

The system was up-scaled and installed at a public washroom (approximately 100 users/day) as a secondary treatment system to treat the partially digested effluents. In the system, the raw sewage first collected in a septic tank and secondly, flows into a soakage pit. However, the soakage pit often overflows due to overloading, thus, the overflowing wastewater pipeline was connected to the ICFiWet system. All components of the up-scaled ICFiWet system was an up-scale of the pilot design. The system was up-scaled to capacity of 15 m^3 to treat 5 m^3 of daily inflow by maintaining 3 days of hydraulic retention time in ICFiWet system. The system was having 3.0 m diameter and 2.8 m depth which included 0.6 m height anaerobic compartment,

Fig. 2 Construction of the up-scaled ICFiWet system

2.0 m height charcoal granular bed, and 0.2 m deep treatment wetland. The up-scaled reactor was built using masonry bricks, reinforced concrete, and finished with cement rendering. Except 0.3 m of freeboard of the wetland, entire system was constructed underground (Fig. 2).

2.3 Water Quality Monitoring

2.3.1 Pilot-Scale System Monitoring

The pilot-scale Integrated Charcoal Filter Constructed Wetland system performances were monitored for 47 days in order to evaluate grey water treatability. The treatment efficiency of the system was measured in terms of pH, electrical conductivity (EC), total solids (TS), total dissolved solids (TDS) by measuring the water quality at the inflow, outflow, and different layers; sampling point 1 (S1) in the anaerobic sludge blanket region, sampling point 2 (S2) at the top of the charcoal granular media, and sampling point 3 (S3) within the constructed wetland (Fig. 1). Water quality parameters of BOD_5, nitrate and phosphate were measured for the inflow and outflow during the operational period. Greywater samples were tested for pH by probe method (Thermo Scientific, model Orion 2 star), Total Solids (TS) by oven dry method, EC and Total Dissolved Solids (TDS) by probe method (Thermo Orion, model 145A), biochemical oxygen demand (BOD_5) by Winkler Titration method, nitrate nitrogen and phosphate by colorimetric method (Jenway 6305 UV/Vis Spectrophotometer).

2.3.2 Up-Scaled System Monitoring

Sampling points were established in up-scaled system, similar to the pilot-scale system and preliminary treatment efficiency was measured in terms of pH, EC, TS, and total suspended solids (TSS), Salinity, BOD_5 for all five measuring points for a period of 18 days. Additionally, Nitrate Nitrogen and Phosphate were measured for inflow and outflow.

3 Results and Discussion

3.1 Performances of Pilot-Scale Integrated Charcoal Filter Constructed Wetland System

Table 1 illustrates the water quality of the inflow, outflow, and the three sampling points during the operational period. The average pH value of influent wastewater was in slightly acidic range but improved to neutral pH while flowing upward through the three consecutive treatment systems. The pH improvement was greater in first anaerobic sludge blanket indicating acidogenic and methanogenic transformations. A neutral pH of 6.69 ± 0.29 facilitates the optimum conditions for acid-producing bacteria and methane-producing bacteria. As shown in Fig. 3a, the time-dependent average pH of the outflow effluent was varied from 6.71 to 8.04 (7.3 ± 0.30) which is in the environmentally acceptable pH range (between pH 6 and 8.5 for inland

Table 1 Quality of wastewater at inflow, outflow and sampling points in pilot-scale system ($\pm$standard deviation, $n = 47$ days)

Parameter	Inflow	S1	S2	S3	Outflow
pH	4.63 ± 0.39	6.69 ± 0.29	6.8 ± 0.29	7.58 ± 0.13	7.3 ± 0.30
EC (mS/cm)	0.37 ± 0.11	1.47 ± 0.72	1.47 ± 0.83	3.12 ± 0.91	1.54 ± 0.74
TS (mg/l)	1908 ± 384	1201 ± 46	1185 ± 633	2646 ± 1637	1208 ± 519
TDS (mg/l)	234 ± 59	719 ± 367	727 ± 428	1586 ± 482	735 ± 482

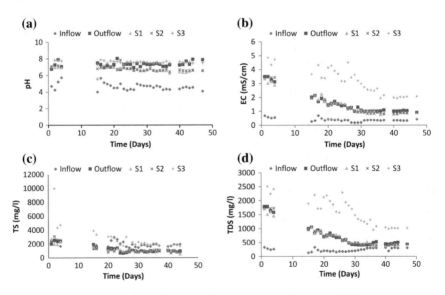

Fig. 3 Variation of **a** pH, **b** EC, **c** TS, and **d** TDS with time in pilot-scale ICFiWet system

surface waters) of treated wastewater before discharging into the natural environment (National Environmental Protection and Quality Regulations, No. 1 of 2008).

The EC variation of different sampling points of the reactor during the study period is shown in Fig. 3b. The artificially prepared greywater (influent) was having low EC compared to outflow waters. The EC was increased through three consecutive treatment systems because of the consumption of complex organic materials by microorganisms and their breakdown within the sludge blanket. Nevertheless, EC within the wetland is relatively high due to the solubility of different ions included in the wetland filter materials (sand, soil, and charcoal). However, with the time EC within the wetland reduced with washing away of impurities and settlement of the filter materials. EC of outflow water is gradually decreased with time from 3.5 to 0.9 mS/cm during the sampling period.

Figure 3c shows the TS variation of the inflow, outflow and within the reactor. As illustrated in Table 1, influent contains 1908 ± 384 mg/l total solids and effluent contains 1208 ± 519 mg/l with a 45.09% of removal efficiency. As shown in Fig. 3d, TDS concentrations of outflow were higher than that of the inflow due to the microbial digestion and breakdown of complex organic substances.

As shown in Fig. 4a, during the first four days of operation, BOD_5 values of the outflow were greater than inflow due to the dissolution of organics in charcoal and wetland bed media substrates. A higher increase of BOD_5 can be observed during 15th day. Formation of microbial flocs due to prolong storage (3 days) of synthetic greywater was the reason for sudden fluctuation of influent waters. But, effluent water quality, particularly BOD_5 was maintained below 90 mg/l with a time-dependent average of 52 ± 33.11 mg/l with approximately 67% of removal efficiency.

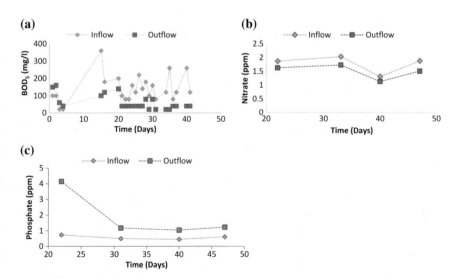

Fig. 4 Variation of **a** BOD_5, **b** Nitrate, and **c** Phosphate concentrations of influent and effluent with time in pilot-scale ICFiWet system

Table 2 Time-dependent averages of the nutrient concentrations and their removal efficiencies

Parameter	Inflow	Outflow	Removal efficiency (%)
BOD$_5$ (mg/l)	156 ± 72.55	52 ± 33.11	66.67
Nitrate (ppm)	1.77 ± 0.28	1.50 ± 0.23	15.3
Phosphate (ppm)	0.58 ± 0.11	1.90 ± 1.3	−205.07

This reduction of BOD$_5$ can occur due to the aerobic and anaerobic degradation of complex organic materials by the microorganisms grown in the filter material surface (Vymazal and Kropfelova 2008; Rani et al. 2011).

Nitrate concentration ranges between 1.31 and 2.03 ppm in the influent wastewater and 1.13–1.73 ppm in the effluent with time-dependent averages of 1.77 ± 0.28 and 1.50 ± 0.23 at influent and effluent, respectively (Table 2). The system shows a 15.3% of nitrate removal efficiency. Nitrate removal is occurred due to biological processes such as denitrification which takes place under anoxic conditions. Anoxic conditions are expected to be present at wetlands except plant root zones and near the surface (Cooper et al. 1996).

Figure 4c illustrates the phosphate concentration variation during the experiment. The influent contains of 0.58 ± 0.11 ppm phosphate concentration which increases to 1.90 ± 1.3 ppm at the outflow as shown in Table 2. This is due to the various microbial activities within the reactor, and complex organic compounds are decomposed into simple forms which release phosphate to the system and limited fixing capacity by filter media and plant.

3.2 Performances of the up-Scaled ICFiWet System

Table 3 shows the preliminary data obtained from the up-scaled ICFiWet system by water quality analysis. While flowing through the three main compartments the

Table 3 Variation of time-dependent water quality parameters of up-scaled ICFiWet system (±standard deviation, $n = 18$ days)

Parameter	Inflow	S1	S2	S3	Outflow
pH	7.93 ± 0.09	7.81 ± 0.12	7.10 ± 0.16	7.06 ± 0.18	7.53 ± 0.16
EC (mS/cm)	3.12 ± 0.49	3.27 ± 0.36	2.00 ± 0.21	1.83 ± 0.30	2.37 ± 0.40
Salinity (‰)	1.72 ± 0.23	1.77 ± 0.16	1.07 ± 0.09	0.98 ± 0.15	1.27 ± 0.20
TS (g/l)	0.81 ± 0.21	0.82 ± 25	1.06 ± 0.35	1.01 ± 0.36	0.64 ± 0.28
TSS (g/l)	0.74 ± 0.36	0.64 ± 0.30	0.58 ± 0.26	0.40 ± 0.36	0.36 ± 0.23
BOD$_5$ (mg/l)	383 ± 23.5	215 ± 71.6	145 ± 44.7	64 ± 34.3	36 ± 12.0
Nitrate N (ppm)	0.19	–	–	–	3.34
Phosphate (ppm)	9.65	–	–	–	3.71

BOD$_5$ of wastewater is reduced from 383 ± 23.5 mg/l to 36 ± 12.0 mg/l which is near to the standard water quality level for surface discharge.

The pH value shows a slight reduction while flowing through three consecutive compartments and it increases slightly at the outflow. However, the outflow pH is lower than the inflow pH and outflow pH is maintained at environmentally acceptable pH range to discharge to the natural environment. EC and Salinity show a similar variation like pH as shown in Table 3. Total solid content increases at S2 (charcoal granular media) region and S3 (wetland media) region due to mixing of filter material and other organic microorganism complexes with flowing wastewater. TSS is reduced gradually from inflow to the outflow through three consecutive treatment systems due to digestion of organic matter by microorganisms.

As this system was fed with a partially digested effluent from a public washroom system, the contribution of ammonia must be high; however, consequently converted to nitrate through nitrification according to data. The inflow of the up-scaled system contains high amount of phosphate (approximately 9.65 ppm) due to heavy usage of detergents for cleaning of public washroom system.

3.3 Conclusion and Recommendation

The pilot-scale Integrated Charcoal filter constructed wetland system shows 67% of BOD removal and the system reduced nitrate concentration by 15.3%. It was also revealed that the system decomposes the phosphorous complexes in influent leaving an increase in effluent phosphate, demonstrating that charcoal substrate and plants have limited capacity to fix phosphate, but recover nutrient for irrigation water use. Thus, it can be concluded that ICFiWet treated greywater can be used in irrigation and home gardening as it contains high amount of soluble phosphate; however, phosphorus adsorption substrate should be used in ICFiWet system if treated water to be discharged into natural waters. In perspective, Integrated Charcoal filter constructed wetland system can be used as a cost effective, efficient wastewater treatment system which can be constructed with locally available materials. High phosphate in outflow water gives an added advantage if the treated water to be used for irrigation/gardening, thus the system shall be coupled with home garden irrigation with treated water.

References

Albalawneh, A., Chang, T. K., & Alshawabkeh, H. (2017). Greywater treatment by granular filtration system using volcanic tuff and gravel media. *Water Science and Technology, 75*(10), 2331–2341.

Ammari, T. G., Al-Zu'bi, Y., Al-Balawneh, A., Tahhan, R., Al-Dabbas, M., Ta'any, R. A., & Abu-Harb, R. (2014). An evaluation of the re-circulated vertical flow bioreactor to recycle rural greywater for irrigation under arid Mediterranean bioclimate. *Ecological Engineering, 70*, 16–24.

Bdour, A. N., Hamdi, M. R., & Tarawneh, Z. (2009). Perspectives on sustainable wastewater treatment technologies and reuse options in the urban areas of the Mediterranean region. *Desalination, 237*(1–3), 162–174.

Boyjoo, Y., Pareek, V. K., & Ang, M. (2013). A review of greywater characteristics and treatment processes. *Water Science Technology, 67*(7), 1403–1424.

Ciuk, K. S. (2015). *Simulating water and pollutant transport in bark, charcoal and sand filters for greywater treatment.* Department of Energy and Technology, Swedish University of Agricultural Sciences, Uppsala, Sweden.

Cooper, P. F., Job, G. D., Green, M. B., & Shutes, R. B. E. (1996). *Reed beds and constructed wetlands for wastewater treatment.* Swindon, Wiltshire, UK: WRc plc.

Dalahmeh, S. S., Pell, M., Vinneras, B., Hylander, L. D., Oborn, I., & Jonsson, H. (2012). Efficiency of bark, activated charcoal, foam and sand filters in reducing pollutants from greywater. *Water, Air, and Soil Pollution, 223*(7), 3657–3671.

El-Khateeb, M. A., & El-Gohary, F. A. (2003). Combining UASB technology and constructed wetland for domestic wastewater reclamation and reuse. *Water Science and Technology, 3*(4), 201–208.

Gross, A., Shmueli, O., Ronen, Z., & Raveh, E. (2007). Recycled vertical flow constructed wetland (RVFCW)-a novel method of recycling greywater for irrigation in small communities and households. *Chemosphere, 66*(5), 916–923.

Kamal, A. S. M., Goyer, K., Koottatep, T., & Amin, A. T. M. N. (2008). Domestic wastewater management in South and Southeast Asia: The potential benefits of a decentralised approach. *Urban Water Journal, 5*(4), 345–354.

Katukiza, A. Y., Ronteltap, M., Niwagaba, C., Kansiime, F., & Lens, P. (2014). A two-step crushed lava rock filter unit for grey water treatment at household level in an urban slum. *Journal of Environmental Management, 133,* 258–267.

Li, F., Wichmann, K., & Otterpohl, R. (2009). Review of the technological approaches for grey water treatment and reuses. *Science of the Total Environment, 407*(11), 3439–3449.

Li, Z., Gulyas, H., Jahn, M., Gajurel, D. R., & Otterpohl, R. (2004). Greywater treatment by constructed wetlands in combination with TiO_2-based photocatalytic oxidation for suburban and rural areas without sewer system. *Water Science and Technology, 48*(11–12), 101–106.

Lorion, R. (2001). *Constructed wetlands: Passive systems for wastewater treatment.* US EPA Technology Innovation Office: Report.

Manz, D. H., & Eng, P. (2004). New horizons for slow sand filtration. In *Proceedings of the Eleventh Canadian National Conference and Second Policy Forum on Drinking Water and the Biennial Conference of the Federal-Provincial-Territorial Committee on Drinking Water, Promoting Public Health Through Safe Drinking Water* (p. 6), Calgary, Alberta.

Nansubuga, I., Meerburg, F., Banadda, N., Rabaey, K., & Verstraete, W. (2015). A two-stage decentralised system combining high rate activated sludge (HRAS) with alternating charcoal filters (ACF) for treating small community sewage to reusable standards for agriculture. *African Journal of Biotechnology, 14*(7), 593–603.

National Environmental (Protection and Quality) Regulations, No. 1 of 2008, National environment Act No 47. of 1980.

Nolde, E. (2000). Greywater reuse systems for toilet flushing in multi-storey buildings. *Urb Water J, 1*(2000), 275–284.

Parkinson, J., & Tayler, K. (2003). Decentralized wastewater management in peri-urban areas in low-income countries. *Environment and Urbanization, 15*(1), 75–90.

Rani, S. H. C., Din, M., Md, F., Yusof, M., Mohd, B., & Chelliapan, S. (2011). Overview of subsurface constructed wetlands application in tropical climates. *Universal Journal of Environmental Research & Technology, 1,* 103–114.

Travis, M. J., Shafran, A. W., Weisbrod, N., Adar, E., & Gross, A. (2010). Greywater reuse for irrigation: Effect on soil properties. *Science of the Total Environment, 408*(12), 2501–2508.

Vymazal, J., & Kropfelova, L. (2008). *Wastewater treatment in constructed wetlands with horizontal sub-surface flow.* Dordrecht, The Netherlands: Springer Science & Business Media.

A Consolidated Stratagem Towards Defenestration of Coke Oven Wastewater Using Various Advanced Techniques—An Analogous Study

U. Pathak, D. D. Mandal, S. K. Jewrajka, Papita Das Saha, T. Kumar and T. Mandal

Abstract Coke oven wastewater produced from steel industry contains hazardous constituents like phenols, cyanides, ammonia, SCN^-, etc., which needs to be treated before expulsion to the environment. Therefore, pre-treatment of these large volume of wastewater is required for maintaining environmental standards. Simulated coke oven wastewater pertaining to effluent characteristics discharged by industrial sector was synthesized. Initially, treatment of wastewater was performed with ozone which was used in combination with activated carbon (AC) and H_2O_2 to increase the degradation of COD. The maximum degradation achieved in the O_3/AC was 76.8% while in with O_3/H_2O_2, the COD removal was 75.8%. The O_3/AC process was found more acceptable in terms of fast rate of COD degradation, time and economy efficiency suitable for handling large volume of wastewater. However, problems of sludge disposal and process hazards diverted the adoption towards microbial treatment using bacterial strain *Alcaligenes faecalis* JF339228 where phenol was degraded up to 80.88% in 76 h from the coke oven mixture. Due to the high toxicity level of coke oven wastewater, only biological treatment fails to treat them effectively. Combined microbial treatment as well as membrane-based separation process (thin-film composite—reverse osmosis membrane) for wastewater purification was also applied. Thin-film composite (TFC) RO membrane was used to treat this solution at pressure of 200 and 300 psi and at different pH of 5, 7 and 8, respectively. The maximum *quantity of* phenol *removed by TFC RO membrane* at 300 psi pressure and pH of 8

U. Pathak · T. Mandal (✉)
Department of Chemical Engineering, NIT Durgapur, Durgapur, India
e-mail: tamal.mandal@che.nitdgp.ac.in

D. D. Mandal
Department of Biotechnology, NIT Durgapur, Durgapur, India

S. K. Jewrajka
Reverse Osmosis Division, CSIR-CSMCRI, Bhabnagar, Gujarat, India

P. D. Saha
Department of Chemical Engineering, Jadavpur University, Kolkata, India

T. Kumar
Department of Petroleum Engineering, ISM Dhanbad, Dhanbad, Jharkhand, India

© Springer Nature Singapore Pte Ltd. 2020
S. K. Ghosh et al. (eds.), *Recent Trends in Waste Water Treatment and Water Resource Management*, https://doi.org/10.1007/978-981-15-0706-9_6

was 76%. Thus, an amalgamated approach of bioremediation and reverse osmosis was sustainable.

Keywords Coke oven wastewater · Ozone · Activated carbon · Alcaligenes faecalis JF339228 · Reverse osmosis

1 Introduction

With the inflation in demand for coal and coke, consumption of freshwater generates coke oven wastewater which results from various onsite operations like quenching, washing, cooling, processing and purification. Other prominent sectors responsible for discharge include petroleum refineries, explosive and phenolic resin manufacturing units, paint and varnish industries, pharmaceuticals plants, textiles units, organic dye processes, metallurgical operations, etc. The complex nature contains a diverse range of hazardous compounds like phenol, ammonia, cyanide, thiocyanate, oil and grease, etc. However, the concentration varies with the classification of coal and the operating conditions involved in the process. These pollutants find their way to surface and groundwater or to the nearby river and the low-level areas. Particularly, phenolic compounds are easily transmitted to the aqueous ecosystem due to their high solubility and volatile nature (Pal and Kumar 2014).

Suitable treatment strategies for coke oven wastewater include coagulation and flocculation, chlorination, chemical oxidation, adsorption, biological degradation and membrane process. Physico–chemical methods have been widely applied due to their high proficiency but cost involvement and evolution of secondary toxicants or dioxins becomes problematic (Kim et al. 2007; Minhalma and de Pinho 2002; Papadimitriou et al. 2006; Staib and Lant 2007). Biological methods are more economical and environmental friendly but cannot be considered initially to handle such extreme levels of toxicity. Biological treatment has been adopted due to its efficacy of degrading phenol and various pollutants into non-toxic end products with reduced operating costs (Lin and Weber 1992). Bioremediation of phenol eventuates through cleavage in the benzene ring by the enzyme secreted by micro-organism. But these bacteria behave disparately in different phenol concentrations and other process conditions. *Pseudomonas sp.* emanates as one of most popularly applied bacteria for bio attenuation of phenol where the latter acts as carbon source and desired energy for cell growth and metabolism (Bandhyopadhyay et al. 2001). Separation using membranes has also emerged to be a potential technology but the drawbacks of fouling have restricted its applicability at preliminary stage. Both physico–chemical and biological treatment techniques are being applied in full-scale industrial use, with satisfactory reports of abatement. Current trends have diverted the treatment technologies to hybrid process like electro catalytic degradation with simultaneous adsorption, solvent extraction, chemical oxidation or irradiation which poses several problems like high process expenses and production of virulent by-products (Bodalo et al. 2008; Spiker et al. 1992). Chemical oxidation process using Fenton's reagent

Table 1 Characteristic features of four different membranes (Bódalo et al. 2009; López-Muñoz et al. 2010; Kislik 2010)

Membrane process	Membrane pore size (μm)	Applied pressure (kg/cm^2)	Rejected solute size (Å)	Permeates through membrane	Suspended particles
MF	0.2–5	<2	>1000	Water dissolved solutes	Suspended particles
UF	0.02–0.2	1–5	50–1000	Water small molecules	Large molecules
NF	0.002–0.01	5–15	5–75	Water monovalent ions	Multivalent ions
RO	<0.002	20–75	1–10	Water	All solutes

also improves the biodegradability of the wastewater. But the major disadvantage of Fenton's reagent appears to be large consumption of $H_2O_2/FeSO_4$ and generation of sludge (Mandal et al. 2010). Agenson et al. (2003) utilized five different membranes for eradication of alkyl phenols (initial concentration-0.05 mg/L) and achieved removal of 90%. For absolute elimination of complex pollutants, simultaneous advanced oxidation and RO process can be applied. Though membranes have elevated rejecting salt levels, they often demonstrate low rejection for some organic molecules (Kucera 2010). Therefore, their pre-treatment with adsorption processes has been executed. The characteristics of the different membranes are expressed in Table 1.

Kumar et al. (2017) implemented ozone as the oxidizing agent where two sets of experiments—one in combination with activated carbon and the other purged with H_2O_2. In continuation to this study, the eradication of phenol in presence of cyanide and ammonia by an isolated bacterial strain *Alcaligenes faecalis* JF339228 from coke processing wastewater has been attempted in combination with TFC RO membrane.

2 Experimental

2.1 Materials

Phenol, ammonia and cyanide standards were purchased from Merck, India to prepare synthetic coke oven wastewater. For the membrane preparation polysulfone and dimethyl formamide (DMF) were procured. Trimesoyl chloride (TMC, Sigma Aldrich) and *m*-phenylenediamine (MPD, Sigma Alrich) are used for preparing TFC RO membrane. All the solvents and salts were purchased from SRL, India.

2.2 Preparation of Coke Oven Wastewater

Feed solution was prepared containing 600 ppm of phenol, 200 ppm of ammonia and 60 ppm of cyanide. The pH of feed solution was 7. Cyanide and ammonia stock solutions were synthesized from 1000 ppm standard solution following serial dilution technique. Phenol stock solutions were developed by dissolving requisite amount in deionized water.

2.3 Microbial Culture and Acclimatization

Alcaligenes faecalis JF339228 was isolated from the drainage effluent of Durgapur steel plant and maintained on nutrient agar plates/slants at 30 °C and pH 7.4 sub-cultured every 30 days for further usage. Sterilization experiments were carried out with the autoclave. Acclimatisation of the microorganism was executed in minimal media to adapt to the toxic environment for essential utilisation. For *Alcaligenes faecalis* JF339228 temperature required for growth is 32–35 °C. The plates were kept in the incubation chamber for 10–12 h. Similar conditions were also maintained for liquid nutrient media. The bacterial growth curve was studied over 0–76 h. The collected samples were centrifuged at 10,000 rpm for 10 min to separate the cell pellet from the supernatant part. The cell pellet was used for measuring the OD in UV—visible spectrophotometer at wavelength 600 nm. Supernatant part was utilized for phenol measurement at 270 nm.

2.4 Synthesis of TFC RO Membrane

PSf casting solution contained dissolved PSf in DMF (15% w/v) at 70 °C. Semi-automated casting machine was used to prepare PSf membrane supported on nonwoven polyester fabric of thickness 100 μm by phase inversion method. The membrane thickness was 30–40 μ. Temperature and humidity of the casting assembly were maintained between 33–35 °C and 13–14%, respectively. PSf-supported fabric was exposed to a gelation bath at a temperature of 24–25 °C containing sodium lauryl sulphate (0.1% w/v) in aqueous DMF (4% w/v).

To prepare TFC membrane with MPD, the PSf support membrane adhered on glass slide was plunged into MPD (2%, w/v) aqueous solution comprizing of glycerol (0.5% w/v) and DMSO (1% w/v). After removing the soaked membrane, its surface was then rubbed to eliminate bubbles formed during the wetting process. Further the MPD-rinsed membrane was immersed into the TMC (0.125% w/v) solution in hexane that formed thin-film coating over the PSf layer. The final material was subjected to heat treatment at 70 °C for 2 min. The humidity and temperature were retained at 62 ± 7% and 30 ± 3 °C, respectively. The synthesized membrane was washed with water and preserved in aqueous glycerol (10% w/v) solution (Peterson 1993; Kesting 1985; Peterson and Cadotte 1990).

The feed solution was transmitted through these membranes for the removal of phenol from the coke oven mixture at applied pressure of 200 psi and 300 psi, respectively. Each experiment was conducted in triplicate. The permeate flux (J) was calculated as:

$$J = \frac{V}{At}$$

J denotes permeate flux (L m^{-2} h^{-1}) in permeate time, t (h), V signifies permeated water volume (litre, L), A stands for membrane cross-sectional area (m^2). The rejection (R %) was evaluated as:

$$R\% = \left(1 - \frac{C_p}{C_f}\right) \times 100$$

where C_f and C_p are the feed and permeate concentrations, respectively.

2.5 Anatomisation

Scanning electron microscopic analysis was performed with a LEO 1430 VP instrument. Atomic force microscopic measurements were executed in semi-contact mode using a Ntegra Aura, (made by Nt-MDT, Moscow) instrument, involving 'Nova' software for image perusal. Images were captured from different zones of the sample. ATR-IR spectroscopy was conducted using an Agilent Cary 600 series.

3 Results and Discussion

3.1 Treatment of Coke Oven Wastewater Using Ozone in Combination with Activated Carbon and Hydrogen Peroxide

The results obtained by Kumar et al. (2017) depicted that COD removal of 76.79 and 75.8% could be achieved by integrated systems like ozonation conjugated with adsorption by activated carbon and ozonation with hydrogen peroxide. Here, hydroxyl radicals serve a prominent role in eradication of pollutants. Though the cost of ozone consumption is a hindrance, but the catalytic property of activated carbon enhances COD removal. Similarly, incorporation of hydrogen peroxide with ozone augments hydroxyl radical generation which favours the removal of pollutants. The combination of Ozone–activated carbon was more preferable in terms of economy, effective COD removal and less consumption of ozone and reaction time.

3.2 Microbial Treatment of Coke Oven Wastewater

Figure 1 shows the growth of *Alcaligenes faecalis* JF339228 with time in presence and absence of coke oven wastewater along with degradation of phenol (mg/l). Results indicate that with time, the microbial biomass increases up to 60 h in the minimal nutrient media where glucose is the only carbon source. This amount was continuously reduced to replace the medium with phenol, cyanide and ammonia. When development of microorganisms occurs in presence of two substrates or carbon sources, it has a tendency to exhaust the more reliable one that can support maximum growth. Similar observations are noted in this study where the glucose is consumed prior to the phenol. Earlier, Mandal et al. (2013) have reported that about 2100 mg/l of phenol could be degraded completely in 90.8 h incubation time. This study reveals that 80.88% of phenol could be degraded in a time period of 76 h. The current work suggests that *Alcaligenes faecalis* JF339228 have the potency to degrade phenol under the inhibitory effect of ammonia and cyanide through the enzymatic activity phenol hydroxylase (Jiang et al. 2007).

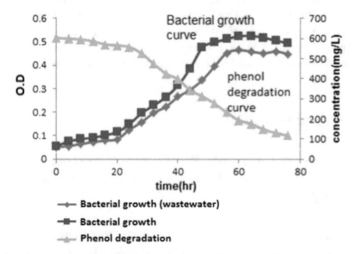

Fig. 1 Bacterial growth curve in absence and presence of coke oven wastewater and phenol degradation curve

3.3 Removal of Phenol from Coke Oven Wastewater by TFC RO Membrane

Initially, the phenol rejection efficiency of the prepared TFC RO membrane was tested with 100, 200, 400 and 600 ppm phenol feed solutions. The pH of the solution was 7. Results depicted from Fig. 2 that the rejection of phenol was almost similar and did not vary marginally with feed concentration. The rejection of phenol was much lower than NaCl rejection capacity of the membrane. Since, TFC RO membrane is negatively charged in nature and the phenol is slightly acidic, the rejection of phenol is governed predominantly by the size exclusion and not by electrostatic repulsion (Srinivasan et al. 2009). Observations also reveal that the concentration of phenol increases with the decrease in permeate flux. At pH 10, the rejection of phenol is higher due to the dissociation nature of phenol into phenolate ions at pH > 8.

From Fig. 3, at an applied pressure of 200 psi, the phenol rejection and the permeate flux by the membrane were lower than 300 psi. The rejection of phenol and flux at pH 8 were higher than that of feed solution of pH 5 and pH 7. Arsuaga et al. (2008) performed experiments with a similar thin-film composite nano-filtration membrane (NF-90) with the phenol feed concentration in the range of 100–500 mg/l. The pH was varied between 2.8 and 5.3 and temperature 20–41 °C. Results revealed that initial organic concentration and operating pressure have a considerable effect on

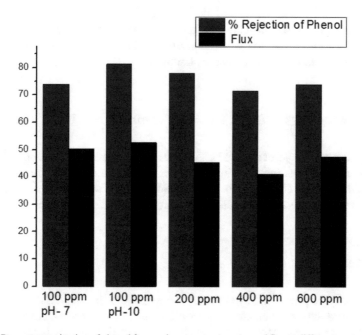

Fig. 2 Percentage rejection of phenol from coke oven wastewater and flux at different concentration

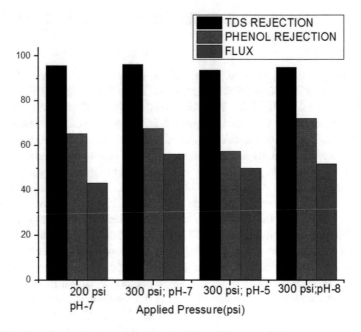

Fig. 3 Rejection of phenol and total dissolved solids at different applied pressure

the permeate flux. On enhancing pressure and deflating organic concentration, the permeate flux increases. However, temperature had no significant influence, but pH has a profound influence on the phenol removal.

3.4 Characterization of TFC RO Membrane

As reported by Jewrajka et al. (2013), TFC RO membrane was synthesized by interfacial polymerization as described in the experimental section. The membrane consists of a typical network-like structure of polyamide which is well reported in the literature with ridges and valleys associated with aromatic C=C stretching of PSf support.

4 Conclusion

Synthetically prepared coke oven wastewater was treated by microbial and TFC RO membrane which was compared to the work reported from ozone–activated carbon treatment. The objective was to remove phenol effectively under the hindering effect of cyanide and ammonia. With the help of coke oven isolated bacterial strain *Alcaligenes faecalis* JF339228, it was found that up to 80.88% of phenol was removed which proved its potency as phenol degrader. On the other hand, with the help of TFC RO membrane at 300 psi and pH 8, phenol can be removed up to 75.25%. The rejection of total dissolved solids was found to be 96.14%. The combination of microbial and TFC RO membrane process will be a novel strategy and greater phenolic removal can be achieved by the combination of these two treatment processes. It may be concluded that only RO process is not as efficient and a combination of the above processes is necessary to accomplish desired abatement.

References

Agenson, K. O., Oh, J. I., & Urase, T. (2003). Retention of a wide variety of organic pollutants by different nanofiltration/reverse osmosis membranes: Controlling parameters of process. *Journal of Membrane Science, 225,* 91–103.

Arsuaga, J. M., López-Muñoz, M. J., Aguado, J., & Sotto, A. (2008). Temperature, pH and concentration effects on retention and transport of organic pollutants across thin-film composite nanofiltration membranes. *Desalination, 221,* 253–258.

Bandhyopadhyay, K., Das, D., Bhattacharya, P., & Maiti, B. R. (2001). Reaction engineering studies on biodegradation of phenol by Pseudomonas Putida MTCC 1194 immobilized on calcium alginate. *Biochemical Engineering Journal, 8,* 179–186.

Bódalo, A., Gómez, E., Hidalgo, A. M., Gómez, M., Murcia, M. D., & López, I. (2009). Nanofiltration membranes to reduce phenol concentration in wastewater. *Desalination, 245,* 680–686.

Bodalo, A., Gomez, J. L., Leon, G., Hidalgo, A. M., & Ruiz, M. A. (2008). Phenol removal from water by hybrid processes: Study of the membrane process step. *Desalination, 223,* 323–329.

Jewrajka, S. K., Reddy, A. V. R., Rana, H. H., Mandal, S., Khullar, S., Haldar, S., et al. (2013). Use of 2,4,6-pyridinetricarboxylicacidchlorideasanovelco-monomer for the preparation of thin film composite polyamide membrane with improved bacterial resistance. *Journal of Membrane Science, 439,* 87–95.

Jiang, Y., Wen, J., Bai, J., Jia, X., & Hu, Z. (2007). Biodegradation of phenol at high initial concentration by *Alcaligenes faecalis. Journal of Hazardous Materials, 147,* 672–676.

Kesting, R. E. (1985). *Synthetic polymeric membranes.* New York: McGraw Hill.

Kim, Y. M., Park, D., Lee, D. S., & Park, J. M. (2007). Instability of biological nitrogen removal in a cokes wastewater treatment facility during summer. *Journal of Hazardous Materials, 141,* 27–32.

Kislik, V. S. (2010). *Liquid membranes: Principles and applications in chemical separations and wastewater treatment* (1st ed.). Amsterdam: Elsevier B.V.

Kucera, J. (2010). *Reverse osmosis: Industrial application and processes.* Hoboken, New Jersey, NJ: Wiley & Sons.

Kumar, A., Sengupta, B., Kannaujiya, M. C., Priyadarshinee, R., Singha, S., Dasguptamandal, D., et al. (2017). Treatment of coke oven wastewater using ozone with hydrogen peroxide and activated carbon. *Desalination and Water Treatment, 69,* 352–365.

Lin, W., & Weber, A. S. (1992). Aerobic biological activated carbon (BAC) treatment of phenolic wastewater. *Environmental Progress, 11,* 145–154.

López-Muñoz, M. J., Arsuaga, J. M., & Sotto, A. (2010). Separation of phenols and their advanced oxidation intermediate products in aqueous solution by NF/RO membranes. *Separation and Purification Technology, 71,* 246–251.

Mandal, S., Bhunia, B., Kumar, A., Dasgupta, D., Mandal, T., Datta, S., & Bhattacharya, P. (2013). A statistical approach for optimization of media components for phenol degradation by Alcaligenes faecalis using Plackett–Burman and response surface methodology. *Desalination and Water Treatment, 51,* 6058–6069.

Mandal, T., Dasgupta, D., Mandal, S., & Datta, S. (2010). Treatment of leather industry wastewater by aerobic biological and Fenton oxidation process. *Journal of Hazardous Materials, 18,* 204–211.

Minhalma, M., & de Pinho, M. N. (2002). Development of nanofiltration/steam stripping sequence for coke plant wastewater treatment. *Desalination, 149,* 95–100.

Pal, P., & Kumar, R. (2014). Treatment of coke wastewater: A critical review for developing sustainable management strategies. *Separation & Purification Reviews, 43,* 89–123.

Papadimitriou, C. A., Dabou, X., Samaras, P., & Sakellaropoulos, G. P. (2006). Coke oven wastewater treatment by two activated sludge systems. *Global NEST Journal, 8,* 16–22.

Peterson, R. J. (1993). *Composite reverse osmosis and nanofiltration membranes* (p. 83). Sci: J. Membr.

Peterson, R. J., & Cadotte, J. E. (1990). Thin film composite reverse osmosis membranes. In M. E. Porter (Ed.), *Handbook of industrial membrane technology* (p. 307). Park Ridger, NJ: Noyes Publication.

Spiker, J. K., Crawford, D. L., & Thiel, E. C. (1992). Oxidation of phenolic and non-phenolic substrates by the lignin peroxidase of Streptomyces viridosporus T7A. *Applied Microbiology and Biotechnology, 37,* 518–523.

Srinivasan, G., Sundaramoorthy, S., & Murthy, D. V. R. (2009). Separation of dimethyl phenol using a spiral-wound RO membrane—Experimental and parameter estimation studies. *Desalination, 243,* 170–181.

Staib, C., & Lant, P. (2007). Thiocyanate degradation during activated sludge treatment of coke-ovens wastewater. *Biochemical Engineering Journal, 34,* 122–130.

Studies on Lead Removal from Simulated Wastewater in Packed Bed Bioreactor Using Attached Growth Technique

Moumita Bose, Siddhartha Datta and Pinaki Bhattacharya

Abstract The present study deals with removal of lead from simulated wastewater using a continuous packed bed bioreactor packed with lead-resistant *Acinetobacter sp 158* immobilized on coconut coir chips. *Acinetobacter sp 158* was isolated from battery industry effluent mud, and biocatalysts were made by using attached growth technique. In order to investigate reactor performance, a set of programmed experiments was performed. The potential of bioconversion of lead through immobilized cell was studied by changing initial substrate concentration along with flow rate with an intention to predict removal of lead through judicious modeling and simulation exercise.

Keywords Bioremediation of lead · Immobilized lead-resistant strain · Biodegradable packing matrix · Packed bed bioreactor

1 Introduction

Amongst the various heavy metals, lead is highly carcinogenic and mutagenic in nature. Random discharge of such hazardous element lead and its compounds as industrial waste from a number of industries, viz battery industry, painting industry, coal-based powder generation industry, mining industry, etc., into the environment (Murthy et al. 2012; Samanta et al. 2012; Vesper et al. 1996; Forstner and Wittmann 2012; Adriano 2001). It is needless to mention that mixing of this heavy metal ions into soil and water beyond its threshold hazardous limit is a potential threat to the entire ecosystem. In order to mitigate lead pollution, chemical industries have been forced to follow lead-containing waste discharge limit set up by Environmental Protective Agencies.

M. Bose (✉) · S. Datta
Chemical Engineering Department, Jadavpur University, Kolkata 700032, India

P. Bhattacharya
Department of Chemical Engineering, Heritage Institute of Technology, Kolkata 700107, India

© Springer Nature Singapore Pte Ltd. 2020   67
S. K. Ghosh et al. (eds.), *Recent Trends in Waste Water Treatment and Water Resource Management*, https://doi.org/10.1007/978-981-15-0706-9_7

The main constrain arises for abating lead contamination is globally applied chemical-based conventional waste treatments reduce the level of lead up to a definite extent which is beyond that the stipulated one. In order to bring down the amount of lead in industrial waste, industry is highly reluctant to adopt again such expensive chemical treatment techniques which make the end product costly. Bioremediation is the only possible route to bring down the amount of lead of industrial effluent near about the threshold value set by Environment Protecting Authorities in developing countries. In order to achieve the permissible limit, conventional waste treatment techniques should be tied up with biological remediation pathways. Although experimental investigation on small-scale batch mode study of lead bioremediation was reported earlier (Chakraborty et al. 2013; Rhee et al. 2014; Li et al. 2016; Kalita and Joshi 2017), a proper programmed and systematic continuous mode operation is needed from bioprocess point of view for successful application of the technology on an industrial scale.

2 Materials and Methods

2.1 Chemicals

Tryptone–Glucose–Yeast Extract Broth (HIMEDIA, India), Lead nitrate [$Pb(NO_3)_2$] (SRL, India), 99.9% of Ethanol (MERCK, India), and distilled water were used for this investigation.

2.2 Preparation of Growth Media

0.1 kg/m^3 lead-incorporated tryptone–glucose–yeast extract broth was used as growth media.

2.3 Preparation Simulated Wastewater

In the present experiment, a stock solution of lead nitrate solution containing kg/m^3 of lead was prepared by adding metered quantity of lead nitrate in double distilled water. In order to make simulated wastewater of definite concentration, stock solution was diluted according to the need.

Fig. 1 Raw coconut chips

2.4 Preparation of Packing Matrix

In the present investigation, outer shell of coconut was collected from local market and cut into small chips (shown in Fig. 1). Coconut chips were washed properly and sundried.

2.5 Microorganism

Previously, isolated lead-resistant bacterium *Acinetobacter sp 158* (Bose et al. 2017) was used in the present investigation.

2.6   Immobilization of the Whole Cell on the Matrix and Its Maintenance

In order to immobilize the whole cell on coconut chips following attached growth technique, 200 ml of culture media and 22.739 gm of coconut chips were introduced in each 500 ml conical flask. These two flasks were autoclaved at 0.2 MPa at 121 °C for 15 min. In order to prepare selective media containing lead metered quantity of UV sterilized lead solution was added into each flask. After that, 20 ml broth culture of lead-resistant *Acinetobacter sp158* was inoculated into each flask in a laminar flow chamber. The open ends of both the flasks were plugged with sterile cotton and placed on rotary shakers by maintaining shaking speed 80 rpm at 37 °C for 48 h. When attached growth was visually appeared, the biocatalyst was loaded into sterile glass column. In order to keep microorganism alive on matrix during shut-down period, the packing portion was submerged in fresh sterile lead supplemented TGY broth.

2.7   Packed Bed Bioreactor Design

A packed bed bioreactor made up of BOROSIL glass column of 615 mm length and 38.8 mm inside diameter was used to treat lead-containing simulated wastewater. In order to support the packing materials, a retaining screen with perforation of 1 mm diameter was inserted at the bottom of the column. In order to collect sample solution from axial direction, the reactor was equipped with three outlet nozzles present 70 mm, 150 mm, and 330 mm, respectively, from the bottom of the column. Simulated wastewater was sent into the packed bed section from the bottom of the column with the help of a peristaltic pump in a predetermined flow rate. In order to maintain aerobic condition, sterile air was sparged into column through air filter after each run. Inlet flow rate of simulated wastewater was measured by rotameter. In order to see overall pressure drop, a U-tube manometer was connected with the packed column. The schematic diagram of bioreactor is shown in Fig. 2. In order to understand the performance of bioreactor, three different operating parameters, viz inlet concentrations of lead (0.05–0.2 kg/m^3), superficial velocities (1.43312–4.29936 m/h), and bed heights (0.07–0.33 m) were considered. After treatment, amount of lead present in simulated wastewater collected from outlet probes was measured using atomic absorption spectrophotometer (Perkin Elmer, Germany).

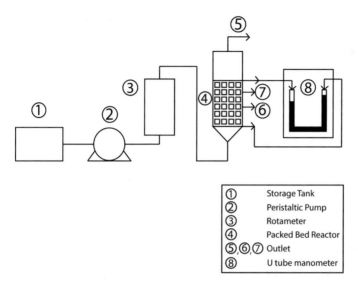

①	Storage Tank
②	Peristaltic Pump
③	Rotameter
④	Packed Bed Reactor
⑤,⑥,⑦	Outlet
⑧	U tube manometer

Fig. 2 Schematic diagram of bioreactor setup

3 Results and Discussions

3.1 Analysis of Performance of Data

In order to explain performance of packed bed bioreactor, a family of curves (Fig. 3a–c) has been constructed using experimental data to determine the lead removal capability of the continuous bioreactor. Figure 3a shows a plot of variation of substrate concentration (S) in axial direction (B) using initial substrate concentration as parameter. It is apparent from those Fig. 3a–c that for each substrate concentration, higher the superficial flow rate, lower is the removal of lead. This is an expected phenomenon, higher superficial velocity gives lower residence time resulting in lower removal of lead, and consequently, lower flow rate gives long residence time to convert higher amount of lead. Another set of plot was drawn (Fig. 4a–c) to understand the relation between substrate concentrations and bed height with superficial velocity as parameter. It is revealed that conversion is gradually increased with the increase of initial substrate concentration for a fixed flow rate due to more availability of electron donor in the vicinity of immobilized cells. In the present set of experiment, the maximum removal of lead was achieved to the tune of 32.4% under the operating condition of 0.2 kg/m^3 initial lead concentration and 1.43312 m/h of superficial substrate flow rate.

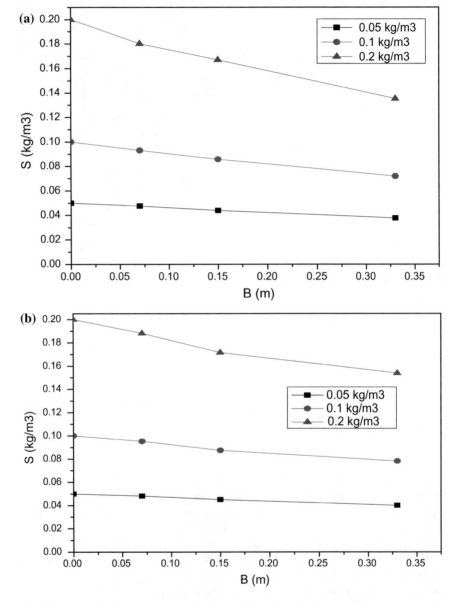

Fig. 3 Variation of substrate concentrations in axial direction using initial substrate concentration as parameter. **a** shows plot for substrate concentration 0.05 kg/m³, **b** shows plot for substrate concentration 0.1 kg/m³, and **c** represents plot for substrate concentration 0.2 kg/m³

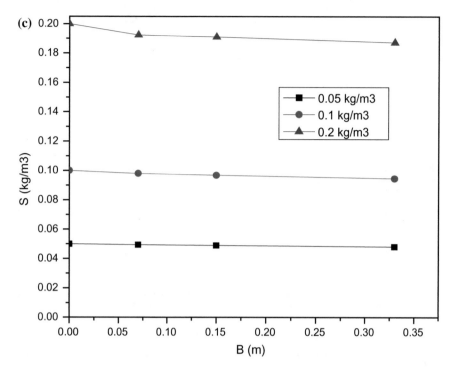

(c)

Fig. 3 (continued)

4 Conclusions

The present investigation reveals successful application of *Acinetobacter sp 158* immobilized form on coconut chips as biocatalyst in a packed bed bioreactor to treat lead-containing wastewater in continuous mode. Easily available coconut chip can be applied as packing matrix to make the process economically viable. The promising experimental results are an indication of carrying out an extensive research using broad spectrum of operating variables.

Acknowledgements One of the authors (Moumita Bose) gratefully acknowledges the University Grants Commissions (UGC), Government of India for providing financial support for this work.

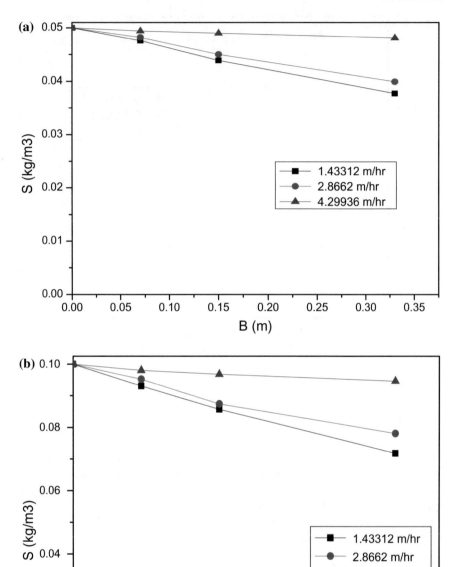

◄**Fig. 4** Variation of substrate concentrations in axial direction using flow rate as parameter. **a** shows plot for flow rate 1.43312 m/h, **b** shows plot for flow rate 2.8662 m/h, and **c** represents plot for flow rate 4.29936 m/h

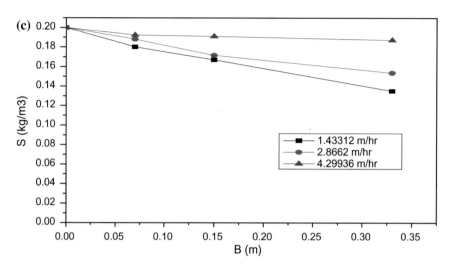

Fig. 4 (continued)

References

Adriano, D. C. (2001). *Trace elements in terrestrial environments: Biogeochemistry, bioavailability and risks of metals* (2nd ed.). Newyork: Springer.

Bose, M., Datta, S., & Bhattacharya, P. (2017). Studies on isolation, characterization and cell growth dynamics of a lead resistant bacterium *Acenetobacter 158*. *Environmental Technology & Innovation, 8,* 103–112.

Chakraborty, S., Mukherjee, A., & Das, T. K. (2013). Biochemical characterization of a lead-tolerant strain of Aspergillus foetidus: An implication of bioremediation of lead from liquid media. *International Biodeterioration & Biodegradation, 84,* 134–142.

Forstner, U., & Wittmann, G. T. U. (2012). *Metal pollution in the aquatic environment* (2nd ed.). Berlin: Springer Science & Business Media.

Kalita, D., & Joshi, S. R. (2017). Study on bioremediation of Lead by exopolysaccharide producing metallophilic bacterium isolated from extreme habitat. *Biotechnology Reports, 16,* 48–57.

Li, X., Peng, W., Jia, Y., Lu, L., & Fan, W. (2016). Bioremediation of lead contaminated soil with Rhodobactersphaeroides. *Chemosphere, 156,* 228–235.

Murthy, S., Bali, G., & Sarangi, S. K. (2012). Biosorption of lead by bacillus cereus isolated from industrial effluents. *British Biotechnology Journal, 2*(2), 73–84.

Rhee, Y. J., Hillier, S., Pendlowski, H., & Gadd, G. M. (2014). Fungal transformation of metallic lead to pyromorphite in liquidmedium. *Chemosphere, 113,* 17–21.

Samanta, A., Bera, P., Khatun, M., Sinha, C., Pal, P., Lalee, A., & Mandal, A. (2012). An investigation on heavy metal tolerance and antibiotic resistance properties of bacterial strain Bacillus sp. isolated from municipal waste. *Journal of Microbiology and Biotechnology Research, 2*(1), 178–179.

Vesper, S. J., Donover-Brand, R., Paris, K. P., & Al-Abed, S. R. (1996). Microbial removal of lead from solid media and soil. *Environmental Science Technology, 32,* 776–781.

Sustainable Growth and Survival of *Litopenaeus Vannamei* Through Wastewater Recycling

K. Kavitha and P. V. Krishna

Abstract Nowadays, aquaculture activities play a major role in increasing the organic waste and toxic compound in the aquatic water bodies. Along with the development of aquaculture, there is an ever-increasing aquaculture waste both on productivity inside the aquaculture system and on the ambient aquatic ecosystem. Therefore, it is apparent that appropriate wastewater recycling processes are needed for sustainable aquaculture development. Hence, the best environmentally acceptable "biofloc technology" (BFT) has been developed. The present study aims the application of BFT to recycle the wastewater nutrients. Pacific white shrimp, Litopenaeus vannamei, was selected to be culture. Three cement tanks C_1, C_2, and C_3 with a stocking density of 120 PL/m^2 were assigned for the study. C_1 is taken as control and C_2, C_3 as test. In BFT, excess of nutrients in aquaculture systems is converted into microbial biomass, which can be consumed by the cultured animals as a food source and the remaining are used for agriculture purpose. This technology has also the capacity to control harmful pathogens in aquaculture. The microorganisms of biofloc successfully converted the toxic ammonia and nitrite into nutritious diet and maintained the water quality in the safety levels for the shrimp. This study has been made to summarize the features and management aspects of the environmental friendly BFT to achieve sustainable aquaculture.

Keywords Biofloc technology · Sustainable aquaculture · Recycling · Wastewater · International society of waste management · Air and water

1 Introduction

Globally, India occupies the second position after China, with sharing 10.0% of world's aquaculture production (FAO 2016). The vast resources in terms of water bodies and species of fish and shellfish in different regions of the country provide a wide array of culture systems and practices. The shrimp, *Litopenaeus vannamei,* is the major contributor in the aquaculture production. However, intensification of

K. Kavitha · P. V. Krishna (✉)
Department of Zoology and Aquaculture, Acharya Nagarjuna University, Nagarjuna Nagar, Guntur, Andhra Pradesh, India

© Springer Nature Singapore Pte Ltd. 2020
S. K. Ghosh et al. (eds.), *Recent Trends in Waste Water Treatment and Water Resource Management*, https://doi.org/10.1007/978-981-15-0706-9_8

the aquaculture activities generates an immense amount of excess organic pollutants that are likely to cause acute toxic effects and long-run environmental risks (Piedrahita 2003). It is reported that effluents released from the aqua farms contain more pollutants than the other industries. A lot of chaos from the public and the environmentalists is due to the discharge of untreated aqua wastages into natural water bodies. Discharge of this untreated water is contaminating the natural water bodies causing outbreak of diseases, imbalancing the biodiversity and ecosystem. To overcome all these challenges and to conserve the ecological sustainability, biofloc technology (BFT) has been developed. Because of economical, environmental, and marketing advantages over the conventional system, BFT has been implemented successfully in shrimp farming. Compared to conventional aquaculture techniques, BFT provides more economical alternative and sustainable technique in terms of minimal water exchange and reduced feed input making it a low-cost sustainable technology for sustainable development (Avnimelech and Kochba 2009; De Schryver et al. 2008). BFT involves the manipulation of C/N ratio to convert toxic nitrogenous wastes into the useful microbial protein and helps in improving water quality under a minimal/zero water exchange system (Ahmad et al. 2017). Moreover, BFT offers a sustainable aquaculture tool by simultaneously addressing its environmental, social, and economical issues. In this context, BFT can also be used in the specific case of maintaining appropriate water quality, and thus, it can reduce the pollution of the pond water. With these positive effects, the present study was undertaken to evaluate the growth and survival of *L. Vannamei* in BFT.

2 Materials and Methods

2.1 Tank and Floc Preparation

Specific-pathogen-free (SPF) seed, *L. vannamei* (PL 9), obtained from a local hatchery. The PL was carried in oxygenated double-layered polythene bags to the laboratory in specific fish transportation tanks. Before releasing the PL, all the tanks were cleaned and disinfected using bleaching powder and dried for 2 days. The biofloc was prepared as per the protocol described by Avnimelech (2012). In brief, on the first day ammonium chloride was added to initiate nitrogen source in the system. On third and fifth day carbon sources were added and on day seven double the quantity of carbon source was added. Due to the addition of carbon and nitrogen source, the color of the water changed to light brown indicating the formation of floc. After tank preparation on the day nine, the PL was brought to farm side and were treated with potassium permanganate, and tank water was added slowly into the seed bags to adjust the salinity and pH; subsequently, the PL was released slowly into the tanks C_1, C_2, C_3 (each with 54.6 m^2) at a stocking density of 120/m^2. Among three tanks (C_1, C_2, C_3), C_1 was taken as control, i.e., without application of biofloc, and C_2, C_3 were taken as test which consist of biofloc. All the tanks were supplied with aeration for sufficient supply of oxygen which is necessary for the formation of biofloc and also for shrimp.

2.2 Analysis of Water Quality Parameters

Water samples for the analysis were carried throughout the study period. Surface water temperature was recorded between 6:00 and 9:00 during the morning hours and between 5:00 and 6:00 during the evening hours with the help of Mercury thermometer. Samples were collected in separate reagent bottles and analyzed by standard methods of (APHA 2005). pH of the water was measured by using the laboratory model ELICO pH meter. Transparency of the water column was assessed with the help of Secchi disk. In case of dissolved oxygen, the collected water samples were fixed with modified/Winkler's titration method (APHA 2005). The cones Imhoff were used to measure the quantity of biofloc which have marked graduations on the outside that can be used to measure the volume of solids that settle from 1 L of system water.

2.3 Microbiological Analysis of Water

Microbiological analysis of water was done regularly thrice a week. To observe the Vibrio count, spread plate technique was used. One ml of cultured water was serial diluted and 10^8, 10^9 ml of serially diluted culture water was spreaded on Thiosulfate citrate bile salts sucrose (TCBS) agar and incubated overnight at 37 °C to find out the presence of *Vibrio* spp.

2.4 Feed Management

L. vannamei fed with commercial diet, Blanca feed pellets (CP Aquaculture, India private limited) was used for four times daily at 6 am, 10 am, 2 pm, 6 pm. The feed quantity was estimated depending on the floc volume. No water exchange was done during the culture period except addition of water to refill the water levels after siphoning out the wastes. Sludge was removed regularly in order to control Nitrite and TSS in the system.

3   Results and Discussions

Water quality parameters, i.e., salinity, temperature, pH, dissolved oxygen, and transparency ranges, are presented in Table 1. Dissolved oxygen (DO) concentration and temperature were similar between treatments throughout the trial. The oxygen dissolved in the system is an important factor not only for the respiration of aquatic organisms but also to maintain favorable chemical and hygienic environment. Muthu (1980) stated that the DO should not be lower than 3.5 ml/l in shrimp culture pond. But in the biofloc system, the aeration should be more than 5.0 mg/l (Avnimelech 2012). In the present investigation, the dissolved oxygen content was ranged from 6.0 to 7.0 mg/l in the system. Continuous aeration was done during the culture period, and therefore, the oxygen level exceeds the limit. The temperature is an ecological factor which influences the hydrological parameters, which in turn influences the metabolism, growth, and other biochemical processes. The optimum range for warm water species is 24–30 °C (Ramanathan et al. 2005). In the present study, the water temperature range varied from 27 to 29 °C. Shrimps are poikilothermic that can modify their body temperature to the environment in normal conditions. The continuous aeration will circulate the equal temperature in the biofloc system. pH is one of the important environmental parameters which decides physiological process of shrimps. The optimum range of pH is from 6.8 to 8.7 and should be maintained for maximum growth and production (Ramanathan et al. 2005). The best condition is from 7.8 to 8.5. If pH changes significantly, it makes shrimp shocked, weakened, and stops eating. If high or low pH extends for a long time, it will make shrimp grow slowly and susceptible to diseases. In the present study, the pH concentration was ranged from 8.0 to 8.5 which is the best condition for shrimp growth. Salinity is the most important factor influencing many functional responses of the organism such as metabolism, growth, migration, osmotic behavior, reproduction. Sudden fluctuation in salinity will affect the osmoregulatory functions of growing organism and lead to mortality. In the present study, the salinity was maintained at 6 to 7 ppt. During the culture period, the salinity was maintained at 6 ppt, but sometimes, it reached 7 ppt. The transparency of water was 15 cm.

The values of ammonia, nitrite, and nitrate were presented in Fig. 1. The results revealed that the ammonia was 0.05 mg/l in the system initially due to the addition of ammonium chloride. After that, it increased to 0.08 mg/l and then gradually decreased. The heterotrophic bacteria present in the biofloc can absorb ammonia 40

Table 1 Water quality parameters	S. no.	Parameters	Range
	1	Salinity (ppt)	6–7
	2	Temperature (°C)	27–29
	3	pH	8.0–8.5
	4	DO (mg/)	6.0–7.0
	5	Transparency (cm)	15–16

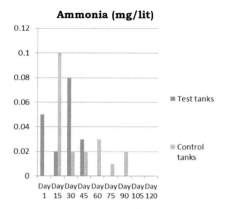

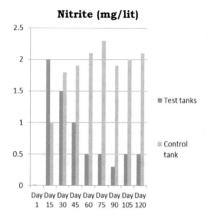

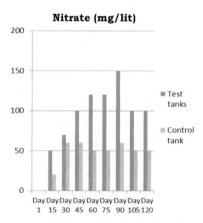

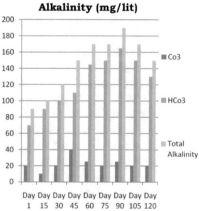

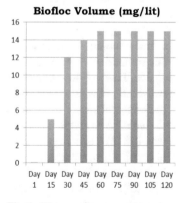

Fig. 1 Water quality parameters

times faster than nitrifying bacteria. Thus, biofloc has the capability of balancing ammonia concentration in the culture system (Ebeling et al. 2006). In addition, De Schryver and Verstraete (2009) stated that biofloc grown in the bioreactor can convert 98% ammonia into nitrate at a concentration of 110 mg L^{-1} day^{-1}. The ability of heterotrophic bacteria to absorb the ammonia is influenced by the ratio of carbon and nitrogen (C/N ratio) in water where the ammonia will be absorbed quickly when C/N was higher (Schneider et al. 2006). The nitrite level was 0.01 mg/l, and then it increased to 2.0 mg/l after 15 days of culture. By increasing the carbon source, removing sludge, and decreasing the feed quantity controlled the nitrite level to 0.5 mg/l. The low value of nitrite in the biofloc system is probably due to the conversion of ammonia to nitrite by heterotrophic bacteria which in turn utilized by micro-algae for growth (Avnimelech 1999; Ebeling et al. 2006). This is an indication that the nitrification process occurred in the culture medium (Stickney 2005). Nitrate, which is the final product from the nitrification of ammonia, is the least harmful nitrogenous compound for shrimp (Van Wyk and Scarpa 1999). In the present study, nitrate ranged between 100 and 150 mg/l. Lower nitrite and higher nitrate concentrations were recorded, due to the formation of biofloc. The total hardness of the water was 1800–1850 mg/l. When alkalinities are maintained from 40 to 160 mg L^{-1} in the BFT system, no significant effect on the TSS, ammonium, nitrites, and nitrates has been determined (Piérri 2012). pH increases when the alkalinity increases above 160 mg L^{-1}, as well as the tendency of increase of heterotrophic and ammonium oxidant bacteria increases with a decrease in nitrites oxidant bacteria (Piérri 2012). Thrice during the culture period alkalinity was reduced to 100 mg/l, in order to increase it sodium bicarbonate was added to maintain the C:N ratio at 15:1. Calcium and magnesium were maintained at a rate of 1:2.5. No H_2S was observed during the entire culture period. In the initial days, the color of the water was light brown and gradually changed to brown color and then to green color. In the day one, the floc volume was 0.1 ml/l, and it increased to 15 ml/l. The quantity and quality of floc were monitored regularly.

The floc volume was presented in Fig. 2. The microscopic observation of floc shows rotifers, copepods, algae, bacteria, etc. But sometimes protozoans like zoothamnium and vorticella were also observed in the floc. Few yellow colonies of Vibrio spp were seen in microbiological analysis; however, there is no disease outbreak throughout the culture period. In aquaculture, the principal source of waste is ultimately the manufactured feeds that are necessary to increase production beyond natural levels (Iwama 1991). According to Avnimelech (2012), reducing the feed quantity and increasing the floc volume reduces the aquaculture waste and also the feed cost. Hence, the adoption of this technology increases the efficiency of feed utilization, because organic and inorganic metabolites, as well as unused or partially used food, are recycled by microorganisms into microalgae and bacterial biomass, which tend to coalesce into flocculated material (bioflocs).

In the present study, biofloc tanks achieved a survival rate of 77% and the growth rate was 27 g with a total production of 135 kgs. Large amounts of manure are required to meet the nitrogen requirements of agricultural crops. For regional agricultural field crops, aquaculture manure is also used as principal nitrogen source for better

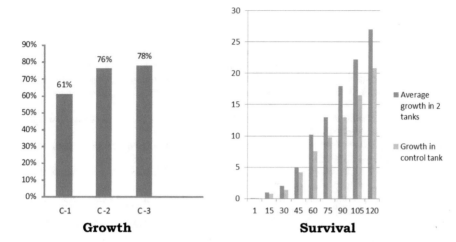

Fig. 2 Growth and survival

effects. On-site, smaller-scale agricultural ventures or non-commercial gardening seem more appropriately scaled for using aquaculture wastes (Yeo et al. 2004). The post-harvested water was used to grow vegetables present near the pond or else the same water can be reused to culture the next crop.

4 Conclusion

Biofloc technology is an environmental and sustainable technology used in aquaculture to maintain water quality through converting nitrogenous waste into bacterial proteinaceous biomass after the addition of carbohydrate sources and also can be subsequently consumed by the cultivated aquatic organisms. In the present study, the biofloc system can maintain the good water quality parameters and recycles the wastewater in shrimp culture. Hence, the BFT is an eco-friendly approach reducing pollution; as there is zero or minimal water exchange to eradicate the environmental pollution and economic problems associated with aquaculture.

Acknowledgements The authors would like to thank the Department of Zoology and Aquaculture, Acharya Nagarjuna University, Guntur, Andhra Pradesh. We would also like to convey our thanks to our beloved friends who gave support and their help in organizing the research work.

References

Ahmad, I., Rani, A. B., Verma, A. K., & Maqsood, M. (2017). Biofloc technology: an emerging avenue in aquatic animal healthcare and nutrition. *Aquaculture International, 25*(3), 1215–1226.

Akpor, O. B., & Muchie, M. (2010). Remediation of heavy metals in drinking water and wastewater treatment systems: Processes and applications. *International Journal of Physical Sciences, 5*(12), 1807–1817.

APHA. (2005). *Standard methods for the examination of the water and wastewater.* Washington, D.C.: American Public Health Association.

Avnimelech, Y. (1999). Carbon/nitrogen ratio as a control element in aquaculture systems. *Aquaculture, 176*(3–4), 227–235.

Avnimelech, Y. (2012). *Biofloc technology, a practical guide book* (2nd ed.). Baton Rouge, Louisiana, EUA: The world Aquaculture Society.

Avnimelech, Y., & Kochba, M. (2009). Evaluation of nitrogen uptake and excretion by tilapia in bio floc tanks, using 15 N tracing. *Aquaculture, 287*(1–2), 163–168.

De Schryver, P., Crab, R., Defoirdt, T., Boon, N., & Verstraete, W. (2008). The basics of bio-flocs technology: The added value for aquaculture. *Aquaculture, 277*(3–4), 125–137.

De Schryver, P., & Verstraete, W. (2009). Nitrogen removal from aquaculture pond water by heterotrophic nitrogen assimilation in lab-scale sequencing batch reactors. *Bioresource Technology, 100*(3), 1162–1167.

Ebeling, J. M., Timmons, M. B., & Bisogni, J. J. (2006). Engineering analysis of the stoichiometry of photoautotrophic, autotrophic, and heterotrophic removal of ammonia–nitrogen in aquaculture systems. *Aquaculture, 257*(1–4), 346–358.

FAO, I. (2016). WFP (2015), the State of Food Insecurity in the World 2015. *Meeting the 2015 international hunger targets: taking stock of uneven progress.* Food and Agriculture Organization Publications, Rome.

Iwama, G. K. (1991). Interactions between aquaculture and the environment. *Critical Reviews in Environmental Science and Technology, 21*(2), 177–216.

Muthu, M. S. (1980, August). Site selection and type of farms for coastal aquaculture of prawns. In *Proceedings of the First Mat. Symposium on Shrimp Farming* (pp. 97–106). Marine Products Export Development Authority, India.

Piedrahita, R. H. (2003). Reducing the potential environmental impact of tank aquaculture effluents through intensification and recirculation. *Aquaculture, 226*(1–4), 35–44.

Piérri, V. (2012). The effect of biofloc technology (BFT) on water quality in white shrimp *Litopenaeus vannamei* culture.

Ramanathan, N., Padmavathy, P., Francis, T., Athithian, S., & Selvaranjitham, N. (2005). *Manual on polyculture of tiger shrimp and carps in freshwater* (pp. 1–161). Tamil Nadu Veterinary and Animal Sciences University, Fisheries College and Research Institute, Thothukudi.

Schneider, O., Sereti, V., Eding, E. H., & Verreth, J. A. (2006). Molasses as C source for heterotrophic bacteria production on solid fish waste. *Aquaculture, 261*(4), 1239–1248.

Stickney, R. R. (2005). *Aquaculture: An introductory text* (p. 265). Massachusetts: CABI Publication.

Van Wyk, P., & Scarpa, J. (1999). Water quality requirements and management. In *Farming marine shrimp in recirculating freshwater systems* (pp. 128–138).

Yeo, S. E., Binkowski, F. P., & Morris, J. E. (2004). *Aquaculture effluents and waste by-products characteristics, potential recovery, and beneficial reuse.*

Reuse of Washing Machine Effluent Using Constructed Wetland: The Circular Economy of Sanitation

Shruti Sharma, B. Lekshmi, Rahul S. Sutar, Yogen J. Parikh, Dilip R. Ranade and Shyam R. Asolekar

Abstract Currently, due to the practice of traditional linear economic model, water resources are under extreme pressure. The lack of technically efficient and economically feasible solutions for water and wastewater treatment and poor management of water assets are responsible for the lack of access to safe water in both rural and peri-urban communities. The present study attempts to achieve the fundamental principles of circular economy which consists of regeneration and restoration by treating the laundry wastewater using constructed wetland and its reuse for various other purposes. The composition of detergents used in a domestic washing machine constitutes of wide range of chemicals. Consequently, the produced laundry wastewater contains varieties of compounds including surfactants, softeners, alkalis, bleach, hydro-tropes, xenobiotic organics, sodium, phosphorus, nitrogen and micro fabrics from clothing. Based on the calculation of water balance on a domestic washing machine, the water consumption has been approximated to be 5000 MLD, while consumption of detergents was estimated to be 3800 TPD (equivalent to 3000 TPD of COD) approximately. Evidently, the treatment of laundry wastewater is the need of the hour. At IIT Bombay, the constructed wetland technology having the horizontal subsurface system entitled 'CW4Reuse' has been developed for treatment and management of effluents. In the present study, 'CW4Reuse' has been utilized for treatment of laundry wastewater from a family consisting of five members (approximately 75 L/d). The wetland bed has a plan area of 1 m^2 and a depth of 0.5 m. *Canna Indica* (family: *Cannaceae*) has been used as the vegetation system for effluent treatment. The qualitative and quantitative analysis of the laundry wastewater showed a significant removal of BOD, COD, TSS, VOC_s heavy metals, nitrogen, phosphorus and pathogens (proxy: MPN). The treated laundry effluent can be reused for gardening, vehicles cleaning and flushing.

Keywords Laundry wastewater treatment · Water reuse · Xenobiotic organics · VOC_s · Constructed wetlands · Surfactants · Detergents

S. Sharma · B. Lekshmi · R. S. Sutar · Y. J. Parikh · D. R. Ranade · S. R. Asolekar (✉)
Centre for Environmental Science and Engineering, Indian Institute of Technology Bombay, Mumbai 400076, India

S. K. Ghosh et al. (eds.), *Recent Trends in Waste Water Treatment and Water Resource Management*, https://doi.org/10.1007/978-981-15-0706-9_9

1 Introduction

Water is a limited natural resource and is essential for the survival of all elements of earth's ecosystem. It plays a crucial role in supporting the day-to-day demands of urban, industrial and agricultural sector and thus holds a special position in a nation's economic growth. This wonder liquid also serves as the basic building block of sustainable development and contributes to the socio-economic growth, healthy environment and survival of living beings. However, in the present situation, due to the ever-growing population, industrialization and other economic actions, the demand as well as usage of freshwater supply has increased many folds. Consequently, it has laid the natural water resources under extreme pressure. For instance, it was found that the consumption of freshwater due to domestic activities contributes up to 70–80% of the total volume of wastewater, and out of domestic usage of freshwater as mentioned above, laundry wastewater constitutes up to 33% (Asano 2002; Harrison 2007; Mohamed et al. 2014a, b).

In the twenty-first century, the issues associated with the water shortage and freshwater supply are of increasing concern worldwide (Fragkou and McEvoy 2016; Hoekstra 2016; dos Santos et al. 2018). Furthermore, the higher carbon and energy footprints as well as high operation and maintenance cost associated with centralized wastewater treatment plants pose a significant challenge towards wastewater reclamation and reuse (Martínez-Huitle et al. 2015). In order to deal with the current situation of water shortage, there is an urgent need of sustainable and strategic planning where the wastewater would be considered as the valuable economic good rather than a waste product. The goals of sustainability can only be achieved by implementing the aforementioned concept of waste to wealth rather than following the traditional linear economic model.

In such scenario, the development of decentralized wastewater treatment systems would be one of the best possible solutions which could be implemented at a small scale with a closed loop of wastewater recycling and reuse facilities (Piratla and Goverdhanam 2015; Capodaglio 2017; Wael et al. 2017). It is a pressing need of today to develop efficient water and wastewater treatment systems that are capable of recycling and reuse of treated effluents (Mekala et al. 2007; Garcia and Pargament 2015). Additionally, the treatment system should be environment friendly, easy to use and should involve low operation and maintenance cost.

In the context of freshwater consumption in domestic activities, washing machines occupy the foremost position. It was estimated that approximately six litres of water are consumed per kg of laundry (dos Santos et al. 2018). It is worthy to note that in the past few decades, the concept of wastewater reclamation and reuse has gained significant attention within the scientific communities, and efforts are being made to recycle and reuse the large volume of effluents coming from various industries such as pharmaceuticals, dairy, leather, agricultural sector and textile industries (Szpyrkowicz et al. 2005; Özcan et al. 2008; Babu et al. 2009; Hajjaji et al. 2013; Rodrigo et al. 2014; Ganiyu et al. 2015; Elabbas et al. 2016; Benazzi et al. 2016; Solano et al. 2013). But, the treatment, recycle and reuse systems dedicated to the laundry effluents are

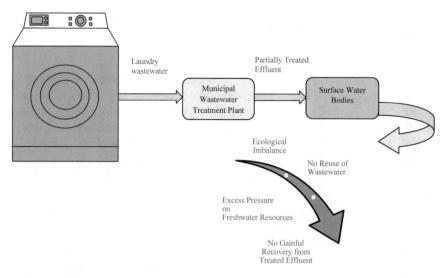

Fig. 1 Current Indian scenario of laundry wastewater

still in the nascent stage, and most often, they are discharged into municipal sewer lines without any prior treatment (Sheth et al. 2017; Durán et al. 2018). In this modern era, nearly every household and industry uses washing machines for cleaning and washing of textiles and in this process generates a huge volume of wastewater which often gets discharged into sewer pipelines despite of having diverse composition which needs special attention before treatment and disposal (Braga and Varesche 2014) Fig. 1 shows the current Indian scenario of laundry wastewater treatment.

Various treatment techniques have been used for treatment of laundry wastewater consisting of physico-chemical methods, oxidation techniques, biological (both aerobic and anaerobic) and integrated methods (Jardak et al. 2016). However, most often such treatment systems are accompanied with high operation and maintenance cost, higher carbon footprint, skilled labours and are energy intensive in nature. Thus, it is imperative to look for sustainable laundry wastewater treatment technologies that would involve low-cost infrastructure and would be easy to operate and maintain without compromising with the water quality. The decentralized wastewater treatment technologies consisting of on-site treatment facilities or sequential treatment systems could be a viable approach for dealing with the heavily loaded washing machine effluents. The natural eco-centric treatment technologies such as constructed wetlands (CWS) would be one of promising solutions for treating the laundry wastewater and the further reuse of treated effluents for high-end purposes.

2   Characterization of Laundry Wastewater and Its Environmental Implications

The laundry wastewater consists of miscellaneous composition and primarily includes surfactants, organic dyes, nitrogen, phosphorus, alkalis, bleach, fillers, suspended solids, micro-fibres, pathogens, micro-plastics and xenobiotic compounds which pose a serious threat to the environment as well as to the living organisms present in the receiving water bodies (Howard et al. 2005; Goel and Kaur 2012; Hidalgo-Ruz et al. 2012; Tripathi et al. 2013; Zavala et al. 2014; dos Santos et al. 2018; Durán et al. 2018). However, the main ingredients of laundry wastewater consist of surfactants, bleach, additives and softeners (Jakobi and Lohr 1987; Abu-Zreig et al. 2003). The quality of laundry effluents depends upon the type of laundered material and the cleaning agents used during washing (Bering et al. 2018). The use of powder detergents often results into significantly higher pH and an elevated concentration of salts like carbonates, bicarbonates, sulphates and chlorides in comparison with liquid detergents (Goel and Kaur 2012). The chemical composition of powder detergent particularly the fillers and builders used during the preparation process might be responsible for such higher pH and concentration of various salts as mentioned above. In a study conducted by Patterson (2004), it was reported that due to the higher solubility of sodium salts and cost-effectiveness, a variety of sodium salts are used in laundry detergents. Some of them include sodium sulphate, sodium tripolyphosphate and sodium carbonate. However, the powder detergents were found to contain exceptionally high content of sodium and phosphorus salts. Primarily, the usage of sodium sulphate as the fillers in laundry powders significantly contributes to the high salinity and electrical conductivity (EC) of laundry wastewater as compared to liquid detergents. The water quality analysis as reported by the author revealed that the usage of powder detergent for laundry purpose resulted into the higher total dissolved solids (TDS), pH, EC, chlorides and sulphates in the laundry effluent. The discharge of such untreated or partially treated laundry effluent negatively affects the soil structural properties as well as the plant growth.

The occurrence of linear alkylbenzene sulphonate (LAS), one of the most widely used anionic surfactants, in water bodies was found to be toxic for aquatic organisms (Spirita Sharmili et al. 2015). The acute toxicity test conducted on the Danio rerio, commonly called as the zebra fish reported that the presence of LAS in the receiving streams causes adverse impacts on the gills and hence arises the respiratory problems in such freshwater organisms. Similar findings have also been reported by Sobrino-Figueroa (2013). In their study, the author revealed the genotoxic characteristics of commercial laundry detergents and anionic surfactants. The environmental impacts occurring due to the discharge of partially treated or treated laundry wastewater have been shown in Fig. 2.

Vasanthi et al. (2013) documented modifications in the biochemical constituents of the fish species *Cirrhinus Mrigala* in the presence of commonly used household detergent known as Tide. Kidney was reported as the most affected organ during the toxicity study. The occurrence of detergents in the natural water bodies adversely

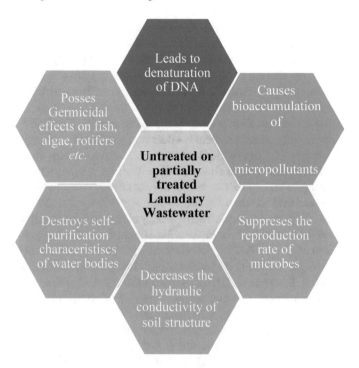

Fig. 2 Environmental implications of untreated, treated or partially treated washing machine effluent

impacts a wide range of aquatic organisms at different trophic hierarchy ranging from micro-algae to macrophytes (Sobrino-Figueroa 2018). The toxicity of the compounds is attributed to the presence of surfactants, enzymes, bleaching agents and silicates.

The toxicity of wheel detergent for mussel (*Lamellidens marginalis*) was determined using the concept of lethal dose 50 (LC 50) and was found that asphyxiation is the common phenomenon of death caused due to the gills impairment (Shingadia and Sakthivel 2003). In connection with the above-reported literature, the acute and chronic toxicity study performed by Saxena et al. (2005) for four different detergents demonstrated that the ill-effects of untreated laundry wastewater for aquatic organisms are not only limited to the gills damage, but it also has the potential to alter the hormonal secretion, reduction in the red blood cells count and surface behaviour.

It has been reported that the anionic surfactants found in laundry wastewater have the potential to bind with peptides and DNA and thus may induce changes in the folding pattern of poly-peptides chain of aquatic organisms present in the downstream, thereby affecting their metabolic functions (Ivanković and Hrenović 2010). With perspectives of human toxicity, it was found that the amphoteric character of anionic surfactants eases their accumulation in living tissues (Cserháti et al. 2002). Similar results have also been reported by Tolls et al. (1994). The fabric softeners used in detergents contain quaternary ammonium complexes as one of the major components, and it is noteworthy that these compounds are potential germicides as well as toxic even at a concentration of mg/L for different aquatic lives such as fish, mussels, rotifers and algae (Lawrence 1970; Waters 1982; Di Nica et al. 2017). The study conducted by Effendi et al. (2017) revealed the suppressing effect of detergents on the bacterial growth in the marine environment. The occurrence of linear alkylbenzene sulphonate in water bodies destroys the organoleptic characteristics and hinders the self-purification process of receiving streams (Cserháti et al. 2002).

In many developing countries, there is a common practice of discharge of laundry wastewater into the soil system which leads to various adverse effects on soil structure and its various physico-chemical characteristics. The highly saline nature of laundry wastewater owing to the substantial use of detergents decreases the hydraulic conductivity of the soil system when discharged untreated (Gross et al. 2008). It has also been found that the composition and permeability of soil also get negatively affected when irrigated with washing machines effluent containing high levels of sodium ions (Mohamed et al. 2018). The high concentration of alkaline laundry detergent increases the soil pH, thus affecting the biological functions of soil micro-organisms (Sivongxay 2005; Travis et al. 2010). Furthermore, it has also been demonstrated that the alkaline laundry wastewater causes dispersion of oil particles and ultimately increases its cation exchange capacity (Anwar 2011).

In addition to above, the occurrence of micro-plastics as well as fibres in the aquatic ecosystem shed during the washing and cleaning process has become one of the worldwide issues nowadays (Cole et al. 2011; Kang et al. 2018a, b). It was approximated that greater than 1900 fibres get discharged off from a single cloth in a washing cycle, thus representing the potential of laundry effluents to be the major source of micro-plastics pollution (Andrady 2011). The ill-effects of micro-plastics on marine flora and fauna are well documented in the various studies and currently are of a serious concern around the globe (Browne et al. 2008, 2011).

3   Constructed Wetlands: The Natural Eco-Centric Treatment Technology

The wetland system that has been designed and constructed aiming at the treatment of water and wastewater by imitating the natural processes within the controlled environment is referred as the constructed wetland (CW) (Vymazal 2011a, b). There are various advantages of CW which include low costs, easy operation and maintenance, low energy consumption, robust, highly efficient for a wide range of contaminants, sustainable, green technology and high ecological value (Carvalho et al. 2017). The few advantages of constructed wetlands have been shown in Fig. 3. Due to these unique characteristics of this eco-centric technology, appreciable progress has been observed in the application of CWs in last few years (Vymazal 2014; Wu et al. 2014; Fu et al. 2016).

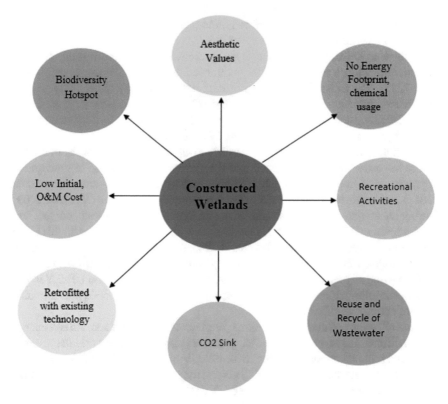

Fig. 3 Advantages of constructed wetlands owing to their application for water and wastewater treatment

The recent advancements in these engineered system have made CWs a promising alternative to the conventional treatment systems for providing access to clean water and sanitation to small communities, households, schools, colleges, etc. (Garcia et al. 2010). Furthermore, it can also be implemented as a decentralized treatment systems in the regions that are deprived of centralized public sewage treatment systems and are economically developing (Brix 1999; Vymazal 2009). Constructed wetlands are suitable for small growing communities (<200,000) and for medium size population (200,000–500,000). Globally, it has been found that constructed wetland can be successfully used for providing efficient treatment to wide varieties of wastewater ranging from urban wastewater to landfill leachates and from municipal to industrial wastewater (Doherty et al. 2015).

Constructed wetlands can be classified on various basis; however, the prime criteria for categorization are hydrology (surface flow and subsurface flow), the type of vegetation growth (emerged, submerged and floating) and flow direction (vertical and horizontal) (Vymazal 2011a, b). Constructed wetlands with surface flow are also commonly known as free surface CWs. It consists of open water surface having floating, submerged or emerged vegetation (Kadlec 2009). The various processes consisting of both physico-chemical and biological were found to be responsible for treatment of wastewater in surface flow CWs (Merz 2000).

In these systems, generally, the organic removal was carried out by microbial degradation, while suspended solids get removed due to physical processes (Kadlec et al. 2017). The satisfactory removal of nutrients particularly phosphorus can also be achieved by using surface flow CWs but at lower removal rates (Vymazal et al. 1998). Mostly, such surface flow CWs are used for providing tertiary treatment to storm water runoff and municipal sewage (Vymazal 2011a, b).

The horizontal subsurface flow constructed wetlands (HFCW) consist of an inlet where the wastewater is introduced into the system and is allowed to flow quietly approximately on a horizontal path through the porous medium under the wetland beds planted with the emergent vegetation to the outlet of the system (Vymazal 2009). During the course of wastewater movement under the beds, it comes in contact with the complex matrix of an anoxic, anaerobic and aerobic zone (Vymazal 2011a, b). The roots and rhizomes provide oxygen to the substrate material and constitute the aerobic zone (Brix 1987). In this case, the system structure is wrapped with liners in order to avoid any leakage (Vymazal and Kropfelová 2008a, b). The HF CW is very effective in elimination of organic, suspended solids, heavy metals and microbial contamination (Vymazal 2009).

Generally, the pollutant degradation is carried out under both aerobic and anaerobic conditions. However, the anaerobic processes play a prominent role in decomposition and removal of contaminants (Brix 1990; Vymazal and Kropfelová 2008a, b). The fundamental advantage of HWCWs is that they are capable of treating the diluted influents derived from combined sewage (Vymazal 2011a, b). The flow within the porous media is sustained by the gravity, and hence, no electrical power is required to maintain the desired flow.

Another important component of a wetland system is substrate which is also known as support matrix (Yang et al. 2018). The substrates are well known for their crucial role in enhancing the treating efficiency of the wetland system. Some of them include providing surfaces for bio-film development, support for the growth of macrophytes, adsorbent for contaminants elimination, etc. (Wu et al. 2015). Apart from conventional pollutants, they are also acknowledged for the exclusion of emerging pollutants like pharmaceuticals, xenobiotics and toxic metals (Dordio and Carvalho 2013; Hu et al. 2015). The traditional substrates, like different types of soil, gravel, were used as the supportive system for the vegetation growth with low or negligible functions on pollutant removal (Wen-Ling et al. 2011). Furthermore, the low removal efficiency and frequent clogging of the traditional substrate have led to the development of new matrix material like construction wastes, clay aggregates and waste tire fragments (Scholz and Xu 2002; Calheiros et al. 2008; Mateus et al. 2012; Chyan et al. 2013; Shi et al. 2017). The vegetation system is one of the unique characteristics of constructed wetlands, and it distinguishes this eco-centric technology from other traditional treatment systems like soil filters, oxidation ponds and aeration channels (Vymazal 2011a, b). These are the essential component of wetland systems owing to their numerous properties in relation to treatment processes and efficiency (Brix

1997). Some of the most commonly used plant species are *Phragmites Australis, Phalaris Arundinacea, Canna Indica* and *Typha Latifolia* (Vymazal 2011a, b).

4　The Potential Role of Constructed Wetlands for Treatment of Laundry Wastewater

The ability of CWs for the treatment of wide range of wastewater wetlands with diverse composition is well evident in past studies. Apart from municipal wastewater (Coleman et al. 2001; Brix and Arias 2005), it could also be used for treating and recycling of wastewater coming from various industries like petrochemical (Yang and Hu 2005), textile (Bulc and Ojstršek 2008), tannery (Calheiros et al. 2007), mining (Gerth et al. 2006), dairy (Mantovi et al. 2003), laundry (Davison et al. 2006), distillery (Billore et al. 2001), etc. In addition to above mentioned, wetland systems are also found to be capable of treating of extensive range of pharmaceuticals (Matamoros et al. 2007; Li et al. 2014), endocrine disruptors (Chapman 2003; Wirasnita et al. 2018), nutrients (Kang et al. 2018a, b; Ndulini et al. 2018), pathogens (Gersberg et al. 1989; Karim et al. 2004) and organic micropollutants (Matamoros et al. 2008; Carranza-Diaz et al. 2014).

In spite of the proven potential, the role of CWs for treatment, recycling and reuse of laundry effluents is still unexplored. As per author's knowledge, the fate, behaviour and transformation of laundry pollutants using dedicated constructed wetlands have not been studied till yet. However, there are few studies available where the discrete removal and decomposition of few surfactants (both anionic and cationic) and xenobiotic compounds have been reported (Huang et al. 2004; Šíma and Holcová 2011; Dordio and Carvalho 2013; Vymazal 2014; Thomas et al. 2017; Pérez-López et al. 2018).

The major research gap found in the literature is the lack of appropriate technology for treatment, processing and reuse of laundry effluents. Currently, the most common practice is to discharge the laundry effluents to municipal sewer lines without any prior treatment which ultimately leads to serious environmental issues Thus, there is need of development of sustainable treatment technology for treatment of laundry wastewater aiming at gainful reuse of treated effluents

5　Innovative Approach for Treatment of Washing Machine Effluents Using Constructed Wetlands—A Way Towards Circular Economy

In light of the above-mentioned facts, an attempt has been made to investigate the treatability of laboratory-scale constructed wetland for treatment of washing machine effluents coming from an urban household. The salient advantages of the wetland system include low cost, easy operation and maintenance and no energy requirement.

In the present study, horizontal subsurface flow-based constructed wetland system entitled 'CW4Reuse' at IIT Bombay has been employed for the treatment and management of washing machine effluent. The laundry effluent is generated from a family consisting of five members which are approximately estimated to be 75 L/d. The plan area of the wetland bed was 1 m², whereas the depth of the bed was restricted to 0.5 m. *Canna Indica* (family: *Cannaceae*) has been used as vegetation system for effluent treatment. The physical and chemical analysis carried out in laboratory-scale reactors showed significant removal of BOD, COD, TSS, VOCS heavy metals, nitrogen, phosphorus and pathogens (proxy: MPN). The obtained water quality was found to be consistent with USEPA standards. The treated laundry effluent can be reused for gardening, vehicles cleaning, floor washing and flushing. The current work is in the progress.

6   Conclusions

The laboratory-scale constructed wetland presents a promising solution for treatment of laundry effluents. Furthermore, the effluent water quality parameters are found to be within the permissible limits as demonstrated by the regulatory standards The treated effluents can be further reused for various household activities like vehicles cleaning, gardening, toilet flushing, floor washing, etc. The current study would be helpful in dealing with the issues of water scarcity in urban and peri-urban regions and would also result into the minimization of freshwater consumption.

7   Future Scope of Work

The facets associated with the tertiary treatment of the treated effluents will be investigated in the coming future for recycle and reuse of laundry effluents for high-end applications.

Acknowledgements The authors of this research paper acknowledge RGSTC and IIT Bombay for providing the partial funding to support the ongoing research work.

References

Abu-Zreig, M., Rudra, R. P., & Dickinson, W. T. (2003). Effect of application of surfactants on hydraulic properties of soils. *Biosystems Engineering, 84*(3), 363–372.

Andrady, A. L. (2011). Microplastics in the marine environment. *Marine Pollution Bulletin, 62*(8), 1596–1605.

Anwar, A. (2011). Effect of laundry greywater irrigation on soil properties. *Journal of Environmental Research and Development, 5.*

Asano, T. (2002). Water from (waste) water—the dependable water resource. *Water Science and Technology, 45*(8), 24.

Babu, B. R., Venkatesan, P., Kanimozhi, R., & Basha, C. A. (2009). Removal of pharmaceuticals from wastewater by electrochemical oxidation using cylindrical flow reactor and optimization of treatment conditions. *Journal of Environmental Science and Health Part A, 44*(10), 985–994.

Benazzi, T. L., Di Luccio, M., Dallago, R. M., Steffens, J., Mores, R., Do Nascimento, M. S., et al. (2016). Continuous flow electrocoagulation in the treatment of wastewater from dairy industries. *Water Science and Technology, 73*(6), 1418–1425.

Bering, S., Mazur, J., Tarnowski, K., Janus, M., Mozia, S., & Morawski, A. W. (2018). The application of moving bed bio-reactor (MBBR) in commercial laundry wastewater treatment. *Science of the Total Environment, 627,* 1638–1643.

Billore, S. K., Singh, N., Ram, H. K., Sharma, J. K., Singh, V. P., Nelson, R. M., et al. (2001). Treatment of a molasses based distillery effluent in a constructed wetland in central India. *Water Science and Technology, 44*(11–12), 441.

Braga, J. K., & Varesche, M. B. A. (2014). Commercial laundry water characterisation. *American Journal of Analytical Chemistry, 5*(01), 8.

Brix, H. (1987). Treatment of wastewater in the rhizosphere of wetland plants-the root-zone method. *Water Science and Technology, 19*(1/2), 107–118.

Brix, H. (1990). Gas exchange through the soil-atmosphere interface and through dead culms of *Phragmites australis* in a constructed wetland receiving domestic sewage. *Water Research, 24,* 377–389.

Brix, H. (1997). Do macrophytes play a role in constructed treatment wetlands? *Water Science and Technology, 35*(5), 11–17.

Brix, H. (1999). How 'green'are aquaculture, constructed wetlands and conventional wastewater treatment systems? *Water Science and Technology, 40*(3), 45–50.

Brix, H., & Arias, C. A. (2005). The use of vertical flow constructed wetlands for on-site treatment of domestic wastewater: New Danish guidelines. *Ecological Engineering, 25*(5), 491–500.

Browne, M. A., Crump, P., Niven, S. J., Teuten, E., Tonkin, A., Galloway, T., et al. (2011). Accumulation of microplastic on shorelines woldwide: Sources and sinks. *Environmental Science and Technology, 45*(21), 9175–9179.

Browne, M. A., Dissanayake, A., Galloway, T. S., Lowe, D. M., & Thompson, R. C. (2008). Ingested microscopic plastic translocates to the circulatory system of the mussel, Mytilus edulis (L.). *Environmental Science & Technology, 42*(13), 5026–5031.

Bulc, T. G., & Ojstršek, A. (2008). The use of constructed wetland for dye-rich textile wastewater treatment. *Journal of Hazardous Materials, 155*(1–2), 76–82.

Calheiros, C. S., Rangel, A. O., & Castro, P. M. (2007). Constructed wetland systems vegetated with different plants applied to the treatment of tannery wastewater. *Water Research, 41*(8), 1790–1798.

Calheiros, C. S., Rangel, A. O., & Castro, P. M. (2008). Evaluation of different substrates to support the growth of *Typha latifolia* in constructed wetlands treating tannery wastewater over long-term operation. *Bioresource Technology, 99*(15), 6866–6877.

Capodaglio, A. (2017). Integrated, decentralized wastewater management for resource recovery in rural and peri-urban areas. *Resources, 6*(2), 22.

Carranza-Diaz, O., Schultze-Nobre, L., Moeder, M., Nivala, J., Kuschk, P., & Koeser, H. (2014). Removal of selected organic micropollutants in planted and unplanted pilot-scale horizontal flow constructed wetlands under conditions of high organic load. *Ecological Engineering, 71,* 234–245.

Carvalho, P. N., Arias, C. A., & Brix, H. (2017). Constructed wetlands for water treatment: New developments.

Chapman, H. (2003). Removal of endocrine disruptors by tertiary treatments and constructed wetlands in subtropical Australia. *Water Science and Technology, 47*(9), 151–156.

Chyan, J. M., Senoro, D. B., Lin, C. J., Chen, P. J., & Chen, I. M. (2013). A novel biofilm carrier for pollutant removal in a constructed wetland based on waste rubber tire chips. *International Biodeterioration and Biodegradation, 85,* 638–645.

Cole, M., Lindeque, P., Halsband, C., & Galloway, T. S. (2011). Microplastics as contaminants in the marine environment: A review. *Marine Pollution Bulletin, 62*(12), 2588–2597.

Coleman, J., Hench, K., Garbutt, K., Sexstone, A., Bissonnette, G., & Skousen, J. (2001). Treatment of domestic wastewater by three plant species in constructed wetlands. *Water, Air, and Soil Pollution, 128*(3–4), 283–295.

Cserháti, T., Forgács, E., & Oros, G. (2002). Biological activity and environmental impact of anionic surfactants. *Environment International, 28*(5), 337–348.

Davison, L., Pont, D., Bolton, K., & Headley, T. (2006). Dealing with nitrogen in subtropical Australia: Seven case studies in the diffusion of ecotechnological innovation. *Ecological Engineering, 28*(3), 213–223.

Di Nica, V., Gallet, J., Villa, S., & Mezzanotte, V. (2017). Toxicity of Quaternary Ammonium Compounds (QACs) as single compounds and mixtures to aquatic non-target microorganisms: Experimental data and predictive models. *Ecotoxicology and Environmental Safety, 142,* 567–577.

Doherty, L., Zhao, Y., Zhao, X., Hu, Y., Hao, X., Xu, L., et al. (2015). A review of a recently emerged technology: Constructed wetland–microbial fuel cells. *Water Research, 85,* 38–45.

Dordio, A. V., & Carvalho, A. J. P. (2013). Organic xenobiotics removal in constructed wetlands, with emphasis on the importance of the support matrix. *Journal of Hazardous Materials, 252,* 272–292.

dos Santos, A. J., de Araújo Costa, E. C. T., da Silva, D. R., Garcia-Segura, S., & Martínez-Huitle, C. A. (2018). Electrochemical advanced oxidation processes as decentralized water treatment technologies to remediate domestic washing machine effluents. *Environmental Science and Pollution Research, 25*(7), 7002–7011.

Durán, F. E., de Araújo, D. M., do Nascimento Brito, C., Santos, E. V., Ganiyu, S. O., & Martínez-Huitle, C. A. (2018). Electrochemical technology for the treatment of real washing machine effluent at pre-pilot plant scale by using active and non-active anodes. *Journal of Electroanalytical Chemistry, 818,* 216–222.

Effendi, I., Nedi, S., & Pakpahan, R. (2017). Detergent disposal into our environment and its impact on marine microbes. In *IOP Conference Series: Earth and Environmental Science* (Vol. 97, No. 1, p. 012030). IOP Publishing.

Elabbas, S., Ouazzani, N., Mandi, L., Berrekhis, F., Perdicakis, M., Pontvianne, S., et al. (2016). Treatment of highly concentrated tannery wastewater using electrocoagulation: Influence of the quality of aluminium used for the electrode. *Journal of Hazardous Materials, 319,* 69–77.

Fragkou, M. C., & McEvoy, J. (2016). Trust matters: Why augmenting water supplies via desalination may not overcome perceptual water scarcity. *Desalination, 397,* 1–8.

Fu, G., Yu, T., Ning, K., Guo, Z., & Wong, M. H. (2016). Effects of nitrogen removal microbes and partial nitrification-denitrification in the integrated vertical-flow constructed wetland. *Ecological Engineering, 95,* 83–89.

Ganiyu, S. O., van Hullebusch, E. D., Cretin, M., Esposito, G., & Oturan, M. A. (2015). Coupling of membrane filtration and advanced oxidation processes for removal of pharmaceutical residues: A critical review. *Separation and Purification Technology, 156,* 891–914.

Garcia, J., Rousseau, D. P., Morato, J., Lesage, E. L. S., Matamoros, V., & Bayona, J. M. (2010). Contaminant removal processes in subsurface-flow constructed wetlands: A review. *Critical Reviews in Environmental Science and Technology, 40*(7), 561–661.

Garcia, X., & Pargament, D. (2015). Reusing wastewater to cope with water scarcity: Economic, social and environmental considerations for decision-making. *Resources, Conservation and Recycling, 101,* 154–166.

Gersberg, R. M., Gearheart, R. A., & Ives, M. (1989). Pathogen removal in constructed wetlands. *Constructed Wetlands for Wastewater Treatment: Municipal, Industrial and Agricultural* (pp. 431–445, 5 fig, 4 tab, 42 ref). Chelsea, MI: Lewis Publishers.

Gerth, A., Hebner, A., Kiessig, G., & Zellmer, A. (2006). Passive treatment of minewater at the Schlema-Alberoda site. In *Uranium in the Environment* (pp. 409–414). Berlin: Springer.

Goel, G., & Kaur, S. (2012). A study on chemical contamination of water due to household laundry detergents. *Journal of Human Ecology, 38*(1), 65–69.

Gross, A., Wiel-Shafran, A., Bondarenko, N., & Ronen, Z. (2008). Reliability of small scale greywater treatment systems and the impact of its effluent on soil properties. *International Journal of Environmental Studies, 65*(1), 41–50.

Hajjaji, W., Ganiyu, S. O., Tobaldi, D. M., Andrejkovičová, S., Pullar, R. C., Rocha, F., et al. (2013). Natural Portuguese clayey materials and derived TiO_2-containing composites used for decolouring methylene blue (MB) and orange II (OII) solutions. *Applied Clay Science, 83,* 91–98.

Harrison, R. M. (Ed.). (2007). *Understanding our environment: An introduction to environmental chemistry and pollution.* Royal Society of chemistry.

Hidalgo-Ruz, V., Gutow, L., Thompson, R. C., & Thiel, M. (2012). Microplastics in the marine environment: A review of the methods used for identification and quantification. *Environmental Science and Technology, 46*(6), 3060–3075.

Hoekstra, A. Y. (2016). A critique on the water-scarcity weighted water footprint in LCA. *Ecological Indicators, 66,* 564–573.

Howard, E., Misra, R., Loch, R., & Le-Minh, N. (2005). Laundry grey water potential impact on Toowoomba soils—Final report.

Hu, T., Haynes, R. J., Zhou, Y. F., Boullemant, A., & Chandrawana, I. (2015). Potential for use of industrial waste materials as filter media for removal of Al, Mo, As, V and Ga from alkaline drainage in constructed wetlands–adsorption studies. *Water Research, 71,* 32–41.

Huang, Y., LaTorre, A., Barceló, D., García, J., Aguirre, P., Mujeriego, R., et al. (2004). Factors affecting linear alkylbenzene sulfonates removal in subsurface flow constructed wetlands. *Environmental Science and Technology, 38*(9), 2657–2663.

Ivanković, T., & Hrenović, J. (2010). Surfactants in the environment. *Archives of Industrial Hygiene and Toxicology, 61*(1), 95–110.

Jakobi, G., & Lohr, A. (1987). Theory of the washing process. Detergent and textile washing—Principles and practice.

Jardak, K., Drogui, P., & Daghrir, R. (2016). Surfactants in aquatic and terrestrial environment: occurrence, behavior, and treatment processes. *Environmental Science and Pollution Research, 23*(4), 3195–3216.

Kadlec, R. H. (2009). Comparison of free water and horizontal subsurface treatment wetlands. *Ecological Engineering, 35*(2), 159–174.

Kadlec, R. H., Knight, R., Vymazal, J., Brix, H., Cooper, P., & Haberl, R. (2017). *Constructed wetlands for pollution control.* IWA Publishing.

Kang, H. J., Park, H. J., Kwon, O. K., Lee, W. S., Jeong, D. H., Ju, B. K., & Kwon, J. H. (2018). Occurrence of microplastics in municipal sewage treatment plants: A review. *Environmental Health and Toxicology, 33*(3).

Kang, Y., Xie, H., Zhang, J., Zhao, C., Wang, W., Guo, Y., et al. (2018b). Intensified nutrients removal in constructed wetlands by integrated Tubifex tubifex and mussels: Performance and mechanisms. *Ecotoxicology and Environmental Safety, 162,* 446–453.

Karim, M. R., Manshadi, F. D., Karpiscak, M. M., & Gerba, C. P. (2004). The persistence and removal of enteric pathogens in constructed wetlands. *Water Research, 38*(7), 1831–1837.

Lawrence, C. A. (1970). Germicidal properties of cationic surfactants. In *Cationic surfactants* (pp. 491–526). New York: Marcel Dekker.

Li, Y., Zhu, G., Ng, W. J., & Tan, S. K. (2014). A review on removing pharmaceutical contaminants from wastewater by constructed wetlands: Design, performance and mechanism. *Science of the Total Environment, 468,* 908–932.

Mantovi, P., Marmiroli, M., Maestri, E., Tagliavini, S., Piccinini, S., & Marmiroli, N. (2003). Application of a horizontal subsurface flow constructed wetland on treatment of dairy parlor wastewater. *Bioresource Technology, 88*(2), 85–94.

Martínez-Huitle, C. A., Rodrigo, M. A., Sirés, I., & Scialdone, O. (2015). Single and coupled electrochemical processes and reactors for the abatement of organic water pollutants: A critical review. *Chemical Reviews, 115*(24), 13362–13407.

Matamoros, V., Arias, C., Brix, H., & Bayona, J. M. (2007). Removal of pharmaceuticals and personal care products (PPCPs) from urban wastewater in a pilot vertical flow constructed wetland and a sand filter. *Environmental Science and Technology, 41*(23), 8171–8177.

Matamoros, V., García, J., & Bayona, J. M. (2008). Organic micropollutant removal in a full-scale surface flow constructed wetland fed with secondary effluent. *Water Research, 42*(3), 653–660.

Mateus, D. M., Vaz, M. M., & Pinho, H. J. (2012). Fragmented limestone wastes as a constructed wetland substrate for phosphorus removal. *Ecological Engineering, 41,* 65–69. *Bioresource Technology, 83*(2), 71–79.

Mekala, G. D., Davidson, B., Samad, M., & Boland, A. M. (2007). *Wastewater reuse and recycling systems: A perspective into India and Australia* (Vol. 128). IWMI.

Merz, S. K. (2000). *Guidelines for using free water surface constructed wetlands to treat municipal sewage.* Department of Natural Resources.

Mohamed, R. M. S. R., Chan, C. M., Wurochekke, A. A., & Kassim, A. H. B. M. (2014b). The use of natural filter media added with peat soil for household greywater treatment. *GSTF Journal of Engineering Technology (JET), 2*(4).

Mohamed, R. M., Al-Gheethi, A. A., Noramira, J., Chan, C. M., Hashim, M. A., & Sabariah, M. (2018). Effect of detergents from laundry greywater on soil properties: A preliminary study. *Applied Water Science, 8*(1), 16.

Mohamed, R., Saphira, R. M., Kassim, M., Hashim, A., Anda, M., & Dallas, S. (2014a). The effects of elements mass balance from turf grass irrigated with laundry and bathtub greywater. *International Journal of Applied Environmental Sciences, 9*(4), 2033–2049.

Ndulini, S. F., Sithole, G. M., & Mthembu, M. S. (2018). Investigation of nutrients and faecal coliforms removal in wastewater using a hydroponic system. *Physics and Chemistry of the Earth, Parts A/B/C.*

Özcan, A., Şahin, Y., Koparal, A. S., & Oturan, M. A. (2008). Propham mineralization in aqueous medium by anodic oxidation using boron-doped diamond anode: Influence of experimental parameters on degradation kinetics and mineralization efficiency. *Water Research, 42*(12), 2889–2898.

Patterson, R. A. (2004). A resident's role in minimising nitrogen, phosphorus and salt in domestic wastewater. In *On-Site Wastewater Treatment X, 21–24 March 2004* (p. 1). American Society of Agricultural and Biological Engineers.

Pérez-López, M. E., Arreola-Ortiz, A. E., & Zamora, P. M. (2018). Evaluation of detergent removal in artificial wetlands (biofilters). *Ecological Engineering, 122,* 135–142.

Piratla, K. R., & Goverdhanam, S. (2015). Decentralized water systems for sustainable and reliable supply. *Procedia Engineering, 118,* 720–726.

Rodrigo, M. A., Oturan, N., & Oturan, M. A. (2014). Electrochemically assisted remediation of pesticides in soils and water: A review. *Chemical Reviews, 114*(17), 8720–8745.

Saxena, P., Sharma, S., Suryavathi, V., Grover, R., Soni, P., Kumar, S., et al. (2005). Effect of an acute and chronic toxicity of four commercial detergents on the freshwater fish Gambusia affinis Baird & Gerard. *Journal of Environmental Science & Engineering, 47*(2), 119–124.

Scholz, M., & Xu, J. (2002). Performance comparison of experimental constructed wetlands with different filter media and macrophytes treating industrial wastewater contaminated with lead and copper.

Sheth, K. N., Desai, M. D., Patel, M., Sheth, K. N., Desai, M. D., & Patel, M. (2017). A study on characterization and treatment of laundry effluent. *International Journal, 4*, 50–55.

Shi, X., Fan, J., Zhang, J., & Shen, Y. (2017). Enhanced phosphorus removal in intermittently aerated constructed wetlands filled with various construction wastes. *Environmental Science and Pollution Research, 24*(28), 22524–22534.

Shingadia, H. U., & Sakthivel, V. (2003). Estimation of LC 50 Values four Lamellidens marginalis lamark with detergent wheel a components of domestic sewage. *Journal of Aquatic Biology, 21*(3), 18.

Šíma, J., & Holcová, V. (2011). Removal of nonionic surfactants from wastewater using a constructed wetland. *Chemistry & Biodiversity, 8*(10), 1819–1832.

Sivongxay, A. (2005). Hydraulic properties of Toowoomba soils for laundry water reuse.

Sobrino-Figueroa, A. (2018). Toxic effect of commercial detergents on organisms from different trophic levels. *Environmental Science and Pollution Research, 25*(14), 13283–13291.

Sobrino-Figueroa, A. S. (2013). Evaluation of oxidative stress and genetic damage caused by detergents in the zebrafish Danio rerio (Cyprinidae). *Comparative Biochemistry and Physiology Part A: Molecular & Integrative Physiology, 165*(4), 528–532.

Solano, A. M. S., de Araújo, C. K. C., de Melo, J. V., Peralta-Hernandez, J. M., da Silva, D. R., & Martínez-Huitle, C. A. (2013). Decontamination of real textile industrial effluent by strong oxidant species electrogenerated on diamond electrode: viability and disadvantages of this electrochemical technology. *Applied Catalysis, B: Environmental, 130*, 112–120.

Spirita Sharmili, V., Kanagapan, M., Sam Manohar Das, S., & Avila Varshini, R. (2015). Studies on the toxicity of Alkylbenzene sulphonate to Zebra fish, Danio rerio (Hamilton). *Journal of Entomology and Zoology Studies, 3*(1), 204–207.

Szpyrkowicz, L., Kaul, S. N., Neti, R. N., & Satyanarayan, S. (2005). Influence of anode material on electrochemical oxidation for the treatment of tannery wastewater. *Water Research, 39*(8), 1601–1613.

Thomas, R., Gough, R., & Freeman, C. (2017). Linear alkylbenzene sulfonate (LAS) removal in constructed wetlands: The role of plants in the treatment of a typical pharmaceutical and personal care product. *Ecological Engineering, 106*, 415–422.

Tolls, J., Kloepper-Sams, P., & Sijm, D. T. (1994). Surfactant bioconcentration-a critical review. *Chemosphere, 29*(4), 693–717.

Travis, M. J., Wiel-Shafran, A., Weisbrod, N., Adar, E., & Gross, A. (2010). Greywater reuse for irrigation: Effect on soil properties. *Science of the Total Environment, 408*(12), 2501–2508.

Tripathi, S. K., Tyagi, R., & Nandi, B. K. (2013). Removal of residual surfactants from laundry wastewater: a review. *Journal of Dispersion Science and Technology, 34*(11), 1526–1534.

Vasanthi, J., Binukumari, S., & Saradhamani, N. (2013). A study on the effect of detergent Tide on the biochemical constituents of the fresh water fish, *Cirrhinus mrigala*. *Journal of Pharmacy and Biological Sciences, 8*(5), 19–22.

Vymazal, J. (2009). The use constructed wetlands with horizontal sub-surface flow for various types of wastewater. *Ecological Engineering, 35*(1), 1–17.

Vymazal, J. (2011a). Constructed wetlands for wastewater treatment: Five decades of experience. *Environmental Science and Technology, 45*(1), 61–69.

Vymazal, J. (2011b). Plants used in constructed wetlands with horizontal subsurface flow: A review. *Hydrobiologia, 674*(1), 133–156.

Vymazal, J. (2014). Constructed wetlands for treatment of industrial wastewaters: A review. *Ecological Engineering, 73*, 724–751.

Vymazal, J., & Kropfelová, L. (2008a). *Wastewater treatment in constructed wetlands with horizontal sub-surface flow*. Dordrecht: Springer.

Vymazal, J., & Kropfelová, L. (2008b). Is concentration of dissolved oxygen a good indicator of processes in filtration beds of horizontal flow constructed wetlands? In J. Vymazal (Ed.), *Wastewater treatment, plant dynamics and management*. Dordrecht: Springer Science+Business Media B.V.

Vymazal, J., Brix, H., Cooper, P. F., Haberl, R., Perfler, R., & Laber, J. (1998). Removal mechanisms and types of constructed wetlands. *Constructed Wetlands for Wastewater Treatment in Europe*, 17–66.

Wael, A, H., Memon, F. A., & Savic, D. A. (2017). An integrated model to evaluate water-energy-food nexus at a household scale. *Environmental Modelling & Software, 93,* 366–380.

Waters, J. (1982). Addendum to the paper on "the aquatic toxicology of DSDMAC and its ecological significance". *Tenside Deterg, 19*(3), 177.

Wen-Ling, Z., Li-Hua, C., Ouyang, Y., Cui-Fen, L. O. N. G., & Xiao-Dan, T. A. N. G. (2011). Kinetic adsorption of ammonium nitrogen by substrate materials for constructed wetlands. *Pedosphere, 21*(4), 454–463.

Wirasnita, R., Mori, K., & Toyama, T. (2018). Effect of activated carbon on removal of four phenolic endocrine-disrupting compounds, bisphenol A, bisphenol F, bisphenol S, and 4-tert-butylphenol in constructed wetlands. *Chemosphere, 210,* 717–725.

Wu, H., Zhang, J., Ngo, H. H., Guo, W., Hu, Z., Liang, S., et al. (2015). A review on the sustainability of constructed wetlands for wastewater treatment: Design and operation. *Bioresource Technology, 175,* 594–601.

Wu, S., Kuschk, P., Brix, H., Vymazal, J., & Dong, R. (2014). Development of constructed wetlands in performance intensifications for wastewater treatment: A nitrogen and organic matter targeted review. *Water Research, 57,* 40–55.

Yang, L., & Hu, C. C. (2005). Treatments of oil-refinery and steel-mill wastewaters by mesocosm constructed wetland systems. *Water Science and Technology, 51*(9), 157–164.

Yang, Y., Zhao, Y., Liu, R., & Morgan, D. (2018). Global development of various emerged substrates utilized in constructed wetlands. *Bioresource Technology*.

Zavala, M. Á. L., Pérez, L. B. S., Reynoso-Cuevas, L., & Funamizu, N. (2014). Pre-filtration for enhancing direct membrane filtration of graywater from washing machine discharges. *Ecological Engineering, 64,* 116–119.

Removal of Methylene Blue Dye by Using Lemon Leaf Powder as an Adsorbent

B. Sarath Babu and G. Yamini

Abstract Wastewater from textile industries is a major cause of water pollution in most developing countries. In order to address the issues of water pollution and high cost for treatment processes, the use of an inexpensive and environmentally benign adsorbent has been studied. Dyes accumulate in wastewater could affect aquatic life, human health, and overall ecosystem adversely. In this project, the lemon leaves powder adsorbent is used to remove methylene blue dye. The adsorbent has been investigated in a batch type and design expert software experiments (Box-Behnken method) to evaluate the process parameters are agitation time, adsorbent size, adsorbent dosage, initial concentration, temperature and at a pH of 5.0. The results of experiments show that the maximum percentage removal of methylene blue by using lemon leaves powder is 90.7% found to be in optimum conditions are concentration of dye is 150 mg/L, temperature of the solution is 313 K, adsorbent dosage is 11 g/L. Maximum adsorption capacity (q_{max}) of lemon leaf powder is found to be 19.19 mg/g.

Keywords Lemon leaves · Methylene blue dyes · Concentration of dye solution · Sorbent dosage

1 Introduction

Dyes are the most harmful pollutants from different processing industries—textile, paper, food, plastic, and cosmetic industries. The removal of these dyes from effluents is a major task and it creates environmental problem. Very low concentration of dyes is effected the light penetration in water bodies and photosynthesis. At the same time, some of the dyes act as carcinogenic and mutagenic in nature (Saha 2010). The conventional methods are used for removing dyes and heavy metals from aqueous solution are coagulation, flocculation, membrane separation, and adsorption, etc. (Hymavathi and Prabhakar 2018). These conventional techniques used for dyes removal are expensive, have average efficiency, and are running in sequential steps (Hasan 2008). Earlier, the adsorbent used in adsorption process is activated carbon,

B. Sarath Babu (✉) · G. Yamini
Department of Chemical Engineering, S V University, Tirupati 517502, India

© Springer Nature Singapore Pte Ltd. 2020   101
S. K. Ghosh et al. (eds.), *Recent Trends in Waste Water Treatment and Water Resource Management*, https://doi.org/10.1007/978-981-15-0706-9_10

which is expensive. Different studies have been made on the possibility of adsorbents using mineral sorbents, activated carbon, peat, chitin, rice husk, soy meal hull, and agrowastes (Deepak et al. 2017).

Adsorption is the adhesion of molecules (ions and atoms) to the surface of a solid or liquid or accumulation of ions on the surface of adsorbent. The material that is being adsorbed or concentrated is called adsorbate and the solid which adsorbs adsorbate is called adsorbent. It is also called surface phenomenon. The adsorption operations can be batch mode and/or continuous mode. Depending on the nature of the forces between adsorbate and adsorbent, the adsorption is of two types—physisorption and chemisorptions.

In physisorption, the binding is by weak Vander Waal's forces. Hence, it is called Vander Waal's adsorption. This force is due to the fluctuating dipole moments on the interaction between adsorbate and adsorbent. In chemisorption, the forces of attraction between adsorbate and adsorbent are as strong as chemical bond.

1.1 Methylene Blue

It has been described as "the first fully synthetic drug used in medicine". It was first prepared in 1876 by German chemist Heinrich Caro. The drugs and dyes are worked in the same way of treatment purpose. This use is no longer recommended because side effects such as vomiting, headache, shortness of breath, high blood pressure, red blood cell breakdown, etc. Methylene Blue is used in redox indicator, water testing, biological staining and aquaculture.

In the current study, removal of methylene blue by lemon leaf powder is investigated through batch experimentation at different levels of initial solution pH, dye concentration, and different sizes of adsorbent.

2   Materials and Methods

2.1 Methylene Blue Preparation

An accurately weighed quantity of dye dissolved in distilled water to prepare a stock solution (1000 ppm). Solution used in the experiment for the desired concentration obtained by further dilution. Dye concentration is determined by using absorbance values measured before and after the treatment at 507 nm with UV visible double beam spectrophotometer.

2.2 Lemon Leaf Powder (Adsorbent)

The lemon leaves can be used to treat nerve disorders like insomnia, nervousness, and palpitation. Lemon leaf will serve as a good and better alternative to Valium and other synthetic sedatives *which by the way have side effects. To get maximum result from lemon leaf, take the preparation* daily for at least one month. The lemon leaf is also used for de-worming. The lemon leaves can be blanched and used to infuse teas, wrap seafood or meat and roasted, steamed or grilled.

2.3 Batch Adsorption Process

Adsorption experiments are carried out in a 500 mL of conical flask containing known amount of adsorbent with dye solution. The flask is then shaken using mechanical rotating shaker at 120 rpm. After that the solution is filtered and the residual dye concentration is measured at λ max using UV Visible spectrophotometer.

Batch adsorption studies are performed to study the effects of adsorbent size, contact time, pH, concentration, temperature, and adsorbent dosage. The effects of concentration, temperature, and adsorbent dosage are studied by using design method. The adsorbent (lemon leaf powder) is weighted by using weight balance. The adsorbate (methylene blue) solutions are prepared and maintained pH by using pH meter. After the weighting of adsorbent and preparation of adsorbate solution in conical flask, put that conical flask in incubator shaker for agitation purpose and to maintain parameters such as time, temperature, and rpm. After the collection, the sample is filtered using filter paper. The resulted solution is analyzed by using UV visible double beam spectrophotometer.

The percentage removal and the amount of dye adsorbed are calculated as follows (Hymavathi and Prabhakar 2017):

$$\text{Percentage removal of dye} = ((C_o - C_f)/C_o) \times 100$$

where, C_o is initial concentration (mg/L) and C_f is final dye concentration (mg/L)
Amount of dye adsorbed $= (C_o - C_f) \, V/W$, $V = $ Volume of dye solution (mL),

$$W = \text{Amount of adsorbent(g)}$$

3 Design of Experiments

Factorial experimental design was used to optimize the preparation conditions and dye removal efficiency. RSM designs allow us to estimate interaction and even

Table 1 Levels of different process variables for removal of methylene blue onto lemon leaf powder	Factors	Level of Box-Behnken		
		Low (−1)	Middle (0)	High (+1)
	Initial concentration (mg/L) (A)	50	100	150
	Temperature (K) (B)	293	303	313
	Adsorbent dosage (g/L) (C)	2	11	20

quadratic effects and hence give us the idea of the (local) shape of the response surface under investigation.

The parameters (time, adsorbent size, agitation, and pH) are done by single-step process to get optimum values. These values are attached to each run of design experiment.

Box-Behnken design is used to optimize the number of experiments with maximum efficiency for an RSM problem involving three factors—initial concentration (A), temperature (B), and adsorbent dosage (C) and three levels (high, middle, and low) as given in Table 1. The levels were fixed based on the preliminary experiment trials and also the available literature. The experimental parts are as shown in Figs. 1 and 2.

Fig. 1 Experimental parts weighing balance and pH meter

Fig. 2 Incubator shaker and UV visible double beam spectrophotometer

4 Results and Discussion

4.1 Effect of Agitation Time

The solution pH is adjusted with 0.1 N NaOH and 0.1 N H_2SO_4 in each flask containing 50 mL solution with 50 mg/L of Methylene blue dye. A required amount of 74 μm particle size of lemon leaves powder is added to the contents of the flasks and the flasks are agitated at constant speed and at room temperature, on an orbital shaker. Flasks are withdrawn periodically; the content filtered and final concentration of dye is estimated by using Eq. 1. The effect of agitation time is shown in Fig. 3.

Fig. 3 Effect of agitation time

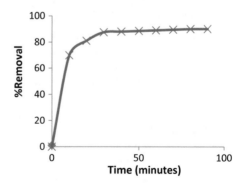

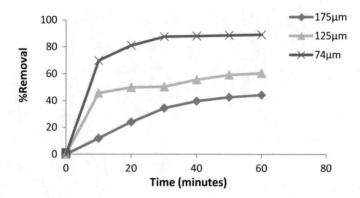

Fig. 4 Effect of adsorbent size

4.2 Effect of Adsorbent Sizes

A 50 mL of the stock solution of 50 ppm taken and with various sizes of adsorbents (175,125 and 74 μm) of adsorbent is taken into 250 mL conical flask, maintaining pH of 5.0, at room temperature and kept for agitation at 120 rpm using an orbital shaker. At the end, the equilibrium time of 80 min samples is withdrawn and filtered using whatman filter paper and tested for its optical density using UV spectrophotometer. The effect of adsorbent sizes is shown in Fig. 4.

4.3 Effect of PH

In order to find the effect of pH, the experiments conducted at various pH values from 2 to 11 in acidic and alkaline conditions, respectively. The pH value is adjusted using 0.1 N NaOH and 0.1 N H_2SO_4. For this work, we have taken 50 mL of 50 ppm of methyl blue dye and 1 g of adsorbent, at room temperature and the sample withdrawn from the shaker at the time of 80 min. The collected sample is tested for its optical densities using the UV visible spectrophotometer. The effect of pH on percentage removal of MB by LLP is shown in Fig. 5.

The experimental and model-predicted responses are given the model (Box-Behnken model) in Table 1. Levels of three process variables for methylene blue removal using lemon leaves are shown in Table 2. The process variables—initial dye concentration, adsorbent dosage, and temperature are studied experimentally, and the complete set of data is fitted to a quadratic model in Eq. 1. Figure 6 % Removal—Predicted versus actual model normal plots for residuals.

$$R1 = 85.80 + 4.43 * A + 1.17 * B - 2.13 * C + 0.29 * AB - 1.70 * AC$$
$$- 0.36 * BC - 0.25 * A^2 - 0.89 * B^2 - 0.3 * C^2 \quad (1)$$

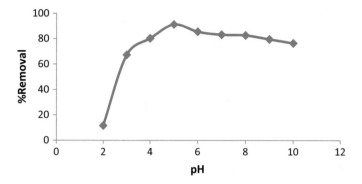

Fig. 5 Effect of pH on percentage removal of MB by LLP

Table 2 Results from Box-Behnken for Methylene blue onto Lemon L

	Sum of squares	df	Mean square	F-value	p-value Prob > F	
Model	221.06	9	24.56	541.40	<0.0001	Significant
A-concentration	157.18	1	157.18	3464.49	<0.0001	
B-temperature	10.97	1	10.97	241.90	<0.0001	
C-dosage	36.25	1	36.25	799.08	<0.0001	
AB	0.32	1	0.32	7.16	0.0317	
AC	11.56	1	11.56	254.81	<0.0001	
BC	0.51	1	0.51	11.27	0.0121	
A^2	0.26	1	0.26	5.63	0.0494	
B^2	3.33	1	3.33	73.31	<0.0001	
C^2	0.39	1	0.39	8.56	0.0221	
Residual	0.32	7	0.045			
Lack of Fit	0.32	3	0.11			
Pure error	0.000	4	0.000			
Cor total	221.38	16				

The R^2 adjusted is 0.996 and R^2 predicted is 0.977; the difference of regression coefficient is <0.2 is well the acceptable norm and the signal. Optimum levels of the three variables studied in the experiment and the model have been presented in Table 3.

There is a good correspondence between the experimental and model-predicted values of the optimum operating conditions.

A comparison of percentage removal of methylene blue—actual versus model-predicted and normal plots for residuals is shown Fig. 6. The residuals, the difference

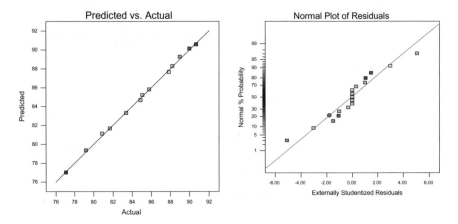

Fig. 6 % Removal—predicted versus actual model normal plots for residuals

Table 3 Optimum levels—methylene blue on to lemon leaves powder

Name of the variable	% Removal of methylene blue	
	Experimental	Optimum values
Dye conc. (Co) mg/L	150	149.63
Sorbent dosage (w), g/L	11	9.6
Temperature (T), K	313	307.5
% Removal	90.7	90.55

between the experimental and observed responses and the predicted responses are examined using the normal probability plots of the residuals. If the model is adequate, the points on the normal probability plots of the residuals should form a straight line.

5 Conclusion

Effectiveness of *Lemon leaves to remove Methylene blue* dyes is thoroughly investigated in the present work and the outcome is promising. Operating parameters, such as concentration of dye solution, initial solution pH, sorbent dosage, and temperature are optimized are shown below.

Optimized conditions	Time (min)	pH	% Removal	q_{max} (mg/g)
Concentration of dye = 150 mg/L	80	5	90.7	19.19
Temperature = 313 K				
Adsorbent dosage = 11 g/L				

Lemon leaves are a good, low-cost, and easily available sorbent for the treatment of methylene blue.

References

Deepak, P., Shikha, S., & Pardeep, S. (2017). Removal of methylene blue by adsorption onto activated carbon developed from *Ficus carica* bast. *Arabian Journal of Chemistry, 10*(1), S1445–S1451.

Hasan, M. B. (2008). *Adsorption of reactive azo dyes on chitosan/oil palm ash composites adsorbent: Batch and continuous studies*. University Sains Malaysia.

Hymavathi, D., & Prabhakar, G. (2017). Studies on the removal of Cobalt(II) from aqueous solutions by adsorption with *Ficus benghalensis* leaf powder through response surface methodology. *Chemical Engineering Communications, 204*(12), 1401–1411.

Hymavathi, D., & Prabhakar, G. (2018). Biosorption of Pb(II) ions onto *Cocos nucifera* leaf powder: Application of response surface methodology. *Environmental Progress & Sustainable Energy, 38*(S1), S118–S127.

Saha, T. K. (2010). Adsorption of methyl orange onto chitosan from aqueous solution. *Water Resource and Protection, 2,* 898–906.

Development of a Low-Cost Column Type Filter Based on Agricultural Waste for Removal of Fluoride from Water

C. M. Vivek Vardhan and M. Srimurali

Abstract Population explosion and subsequent increase in demand for food have resulted in adoption of modern methods of agriculture, thereby generating huge quantities of agricultural wastes. India being an agricultural nation generates large amount of Rice Husk warranting proper utilization before disposal. Concomitantly, fluorosis caused due to consumption of excessive fluoride is posing a serious threat to mankind. Therefore, in this study, an attempt has been made to develop a low-cost column type filter based on Rice Husk for removal of fluoride from water. Lanthanum was used to enhance the fluoride removal capabilities of Rice Husk Ash. Continuous-flow column studies were conducted and investigated for its potential in removing fluoride from water. It was observed that 37.8 L of water of initial fluoride concentration 10 mg/L could be brought down to permissible limits in 1890 min and can be safely consumed. The developed low-cost filter utilizes an agricultural waste and protects people from fluorosis especially of rural and isolated communities.

Keywords Adsorption · Defluoridation · Water · Low-cost · Lanthanum

1 Introduction

Agricultural wastes are generated in huge quantities in agriculture-based nations such as India. It is estimated that the annual generation of Rice Husk in India is about 24 million tons (Pandey et al. 2012). Though a minor portion of it finds application in production of silica, cement and refractory bricks, majority of it is dumped and remains non-utilized. Therefore, an effective way of utilizing the same is required. On the other hand, contamination of groundwater with fluoride and the resulting fluorosis is a major threat faced by mankind. Fluoride has dual impacts on living organisms. In lower concentrations, fluoride is very beneficial as it assists in formation of bones

C. M. Vivek Vardhan (✉)
Department of Civil Engineering, Malla Reddy Engineering College (A), Maisammaguda(H), Medchal, Hyderabad 500100, Telangana State, India

M. Srimurali
Department of Civil Engineering, Sri Venkateswara University College of Engineering, S.V. University, Tirupati, India

© Springer Nature Singapore Pte Ltd. 2020
S. K. Ghosh et al. (eds.), *Recent Trends in Waste Water Treatment and Water Resource Management*, https://doi.org/10.1007/978-981-15-0706-9_11

and prevents dental decay, whereas, in higher concentrations, it causes serious health issues such as skeletal fluorosis, dental fluorosis, mental derangements, dwarfishness, etc. Therefore, it is imperative that fluoride has to be removed from water before consumption. The World Health Organization (WHO) has stipulated a maximum permissible limit of 1.5 mg/L of fluoride in drinking water. Various technologies are available for defluoridation of water and are partially successful to various extents. But, as fluorosis is not yet completely eradicated, the focus has recently shifted to use of rare earth materials as adsorbents for defluoridation with a fairly reasonable degree of success (Vivek Vardhan and Srimurali 2015, 2016, 2018). However, research is still deficient on achieving a dual advantage of simultaneously utilizing waste and removing fluoride from water. Therefore, in this study, an attempt was made to employ Rice Husk Ash impregnated with Lanthanum for removal of fluoride from water in continuous-flow mode and thereby develop a filter using endemically available wastes. Thus, the main objectives of this work are to impregnate Lanthanum on to Rice Husk Ash and to test it for fluoride removal at varying rates of flow rates, initial fluoride concentrations and bed depths.

2 Materials and Methods

2.1 Chemicals

All reagents used in the present investigation were procured from E. Merck Ltd, India and were of analytical reagent grade. Sodium fluoride of 221 mg was dissolved in distilled water and made up to 1 L in a volumetric flask to prepare a stock fluoride solution of 100 mg/L. Stock solution was diluted appropriately to obtain fluoride solutions of required working concentrations.

2.2 Adsorbent

Rice Husk Ash (RHA) was procured from a local rice mill and was sieved to a mean size of 475 μm.

2.3 Adsorbent Preparation

Lanthanum carbonate of varying weights was mixed with 0.05L of distilled water, and 0.1 N HCl was added to it dropwise until Lanthanum Carbonate got dissolved completely. To this solution, 0.1 N NaOH was added under constant stirring until precipitates were visually observed. RHA of mean size 475 μm and 20 ± 0.2 g weight was added to this mixture. This mixture was mixed for 6 h in a magnetic stirrer. Subsequently, it was decanted, dried and heated at 300 °C for 4 h in a muffle furnace. It was thoroughly washed with distilled water and dried. The obtained material is Lanthanum-Impregnated Rice Husk Ash (LIRHA).

2.4 Continuous-Flow Experiments

Fixed bed down-flow column studies were performed using a Pyrex glass column of inner diameter 2.54 cm and depth 1 m. Cotton was placed at the base to support aggregates. The column was filled with aggregate of mean size 600 μm to a depth of 30 ± 1 cm at the base. On its top, LIRHA of varying depths of 20 cm, 30 cm and 40 cm was filled. Aggregate of mean size 600 μm and of varying depths of 30 ± 1 cm, 40 ± 1 cm and 50 ± 1 cm was filled on the adsorbent to ensure an equally distributed flow of tap water spiked fluoride solution onto the sorbent bed. The set-up of the column is depicted in Fig. 1. Aqueous fluoride solutions of various concentrations of 5 ± 0.1 mg/L, 10 ± 0.1 mg/L and 15 ± 0.1 mg/L were passed

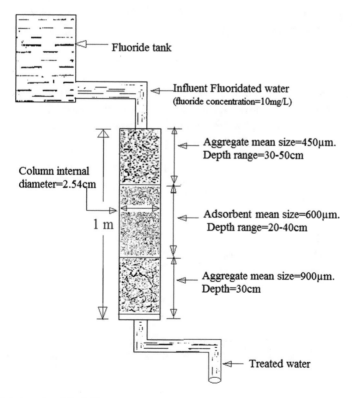

Fig. 1 Column set-up block diagram

through it. Flow rates of the solution were varied at 10 ± 0.2 ml/min, 15 ± 0.2 ml/min and 20 ± 0.2 ml/min. At regular intervals of time, effluent samples were collected and analysed for residual fluoride concentration, until the effluent concentration of fluoride equalled the influent concentration.

2.5   Analysis of Continuous Sorption Experiments

Analysis of breakthrough curves gives an insight into the performance of fixed-bed column. The time taken for breakthrough and time taken for complete exhaustion of column gives characteristic information for determining the performance and dynamics of adsorption of a column (Chen et al. 2013). Breakthrough of the column is considered to have occurred when the effluent concentration (C_t) has reached about 0.1% of influent concentration (C_o), and exhaustion of column is considered to have occurred when effluent concentration has reached about 95% of influent concentration (Kundu and Gupta 2007). However, from fluoride removal point of view, 15% fluoride removal (i.e. up to effluent concentration of 1.5 mgF$^-$/L) is considered for analysis. A breakthrough curve is obtained by plotting values of (C_t/C_o) versus (t) (Han et al. 2009). Effective effluent volume (V_{eff}) can be calculated from Eq. (1) (Ramavandi et al. 2014).

$$V_{eff} = Q \cdot t_{total} \tag{1}$$

where Q is volumetric flow rate (ml/min), and t_{total} is flow time (min). The value of total mass of fluoride adsorbed (q_{total}) (mg) can be obtained from the area under the breakthrough curve, found by integrating the adsorbed concentration (C_{ad}, mg/L), within the functioning time limits (Ramavandi et al. 2014) (Eq. 2).

$$q_{total} = \frac{Q}{1000} \int_{t=0}^{t=total} C_{ad} dt \tag{2}$$

where $C_{ad} = C_o - C_t$.

The maximum fluoride adsorption capacity (q_{eq}) can be calculated as follows, as given in Eq. (3)

$$q_{eq} = \frac{q_{total}}{m} \tag{3}$$

where m is dry weight of adsorbent used in column (g).

Fluoride was analysed using SPADNS method (APHA 2005), using a spectrophotometer, Evolution 201.

3 Results and Discussion

3.1 Continuous-Flow Column Studies

To evaluate the practical applicability of LIRHA for defluoridation, in continuous-flow mode, fluoride-rich water was passed through columns consisting of varying depths, at different flow rates and different influent fluoride concentrations.

3.1.1 Effect of Bed Depth

LIRHA bed depths were varied, and the corresponding variations in related parameters such as time for exhaustion of column and effective volume of water treated were evaluated and presented in Table 1. Increase in bed depth resulted in increase in the effective volume of water treated and took more time for the column to get exhausted. Breakthrough curves were drawn between t versus (C_t/C_o) for various bed depths, and the plots are presented in Fig. 2. It is clear from this figure that with

Table 1 Values of column parameters obtained at different bed depths, flow rates and initial fluoride concentrations for defluoridation using LIRHA

Q (ml/min)	(C_o) (mg/L)	Z (cm)	t_{total} (=Total flow time in min)	V_{eff} (ml)	q_{total} (mg)	q_o (mg/g)
15	10	20	2790	41,850	273.1	4.149
15	10	30	5130	76,950	468	4.739
15	10	40	6930	103,950	770.3	5.85
10	10	30	12,780	127,800	783	7.928
20	10	30	1890	37,800	270.1	2.735
15	5	30	11,700	175,500	484.3	4.904
15	15	30	2790	41,850	343.1	3.475

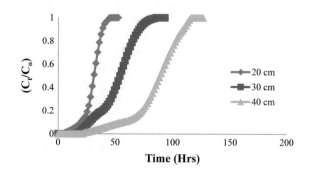

Fig. 2 Breakthrough curves for the effect of bed depth on defluoridation of water using LIRHA

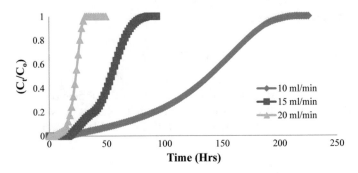

Fig. 3 Breakthrough curves for the effect of flow rate on defluoridation of water using LIRHA

increasing bed depth, the slope of column gradually decreases, marking a broadening of the mass transfer zone, resulting in increased fluoride sorption.

3.1.2   Effect of Flow Rate

Rate of flow of influent fluoride solution into the column was varied, and the corresponding variation of dependent parameters is presented in Table 1. It can be observed from this table that with increase in flow rate, the column gets exhausted rapidly because it gets saturated rapidly. Figure 3 shows the breakthrough curve drawn between t versus (C_t/C_o) for varying flow rates. From this figure, it can be seen that, with increasing flow rate, the breakthrough curves become steeper, and the uptake of fluoride too decreases. This could be due to adsorption taking place, only through film diffusion, because of lesser residence time (Ye et al. 2018).

3.1.3   Effect of Initial Fluoride Concentration

Concentration of fluoride in the influent aqueous solution was varied, and its impact on defluoridation in continuous mode was studied. The influence of variation of initial fluoride concentration on the effective volume treated, and uptake of fluoride and sorption capacities are presented in Table 1. From the table, it can be observed that, with increase in influent fluoride concentration, the adsorption capacity of column too increased gradually. Higher fluoride concentration effectively overcomes mass transfer resistance and therefore transfers fluoride effectively on to the surface of LIRHA. Breakthrough curves were plotted between t versus (C_t/C_o) for different influent fluoride concentrations and are presented in Fig. 4. From this figure, it can be seen that with increase in initial fluoride concentration, the gradient of the breakthrough curve increases, resulting in decreased mass transfer zone. This indicates that intraparticle diffusion controls the adsorption process (Manna et al. 2018).

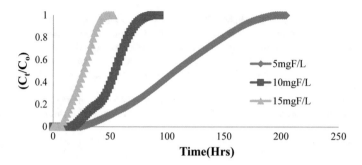

Fig. 4 Breakthrough curves for the effect of initial fluoride concentration on defluoridation of water using LIRHA

4 Concluding Remarks

The aim of this work is to simultaneously utilize the agricultural waste Rice Husk Ash and to remove fluoride from water.

Continuous-flow column studies demonstrated that LIRHA is capable of removing fluoride from water to less than the permissible limits.

Thus, the agricultural waste, Rice Husk can be effectively utilized for removal of fluoride from water, especially in rural and isolated communities.

References

Chen, X., Lu, A., & Qu, G. (2013). Preparation and characterization of foam ceramics from red mud and fly ash using sodium silicate as foaming agent. *Ceramics International, 39*(2), 1923–1929. https://doi.org/10.1016/j.ceramint.2012.08.042.

Han, R., et al. (2009). Adsorption of methylene blue by phoenix tree leaf powder in a fixed-bed column: Experiments and prediction of breakthrough curves. *Desalination, 245*(1–3), 284–297. https://doi.org/10.1016/j.desal.2008.07.013.

Kundu, S., & Gupta, A. K. (2007). As(III) removal from aqueous medium in fixed bed using iron oxide-coated cement (IOCC): Experimental and modeling studies. *Chemical Engineering Journal, 129*(1–3), 123–131.

Manna, S., et al. (2018). Biomass for water defluoridation and current understanding on biosorption mechanisms: A review. *Environmental Progress & Sustainable Energy, 37*(5), 1–13. http://doi.wiley.com/10.1002/ep.12855.

Pandey, R., Sar, S. K., & Bhui, A. K. (2012). Feasibility of installating rice husk power plant in Chhattisgarh to meet sustainable energy demands. *International Journal of Advanced Engineering Research and Studies*. (May 2017).

Ramavandi, B., Farjadfard, S., & Ardjmand, M. (2014). Mitigation of orange II dye from simulated and actual wastewater using bimetallic chitosan particles: Continuous flow fixed-bed reactor. *Journal of Environmental Chemical Engineering, 2*(3), 1776–1784. https://doi.org/10.1016/j.jece.2014.07.023.

Vivek Vardhan, C. M., & Srimurali, M. (2015, March). Defluoridation of drinking water using a novel sorbent: Lanthanum-impregnated green sand. *Desalination and Water Treatment, 3994,* 1–11. http://www.tandfonline.com/doi/abs/10.1080/19443994.2015.1012330.

Vivek Vardhan, C. M., & Srimurali, M. (2016). Removal of fluoride from water using a novel sorbent lanthanum-impregnated bauxite." *SpringerPlus, 5*(1), 1426. http://springerplus.springeropen.com/articles/10.1186/s40064-016-3112-6.

Vivek Vardhan, C., & Srimurali, M. (2018). Preparation of lanthanum impregnated pumice for defluoridation of water: Batch and column experiments. *Journal of Environmental Chemical Engineering.*

Ye, Y., et al. (2018). Fluoride removal from water using a magnesia-pullulan composite in a continuous fixed-bed column. *Journal of Environmental Management, 206,* 929–937.

Oxidative Photocatalytic Degradation of Methylene Blue in Wastewater

Ajit Kumar Tallapaka, Jyothi Thati and Sailu Chintha

Abstract Aqueous methylene blue (MB) dye solution was subjected to photocatalytic degradation by ultraviolet (UV) irradiation with titanium dioxide (TiO_2) catalyst. Batch experimental studies were conducted with 250 ml of different methylene blue concentrations of 20, 30, and 40 mg/l, with catalyst loadings of 25, 50, and 100 mg, at different pH of 3, 7, and 9 to find the degradation of dye. A maximum degradation of 86% methylene blue dye was achieved for concentration 20 mg/l using 100 mg TiO_2 catalyst irradiated for 190 min in presence of UV light of wavelength 254 nm.

Keywords Photocatalysis · Titanium dioxide · Advanced oxidation process · Methylene blue · International society of waste management · Air and water

1 Introduction

In this modern society, any kind of wastewater stream from industrial processes or households contains organic chemicals because of its vast usage, disposal of the same without proper treatment results in exposure to humans and environment (Belgiorno et al. 2007). In textile manufacturing and dyeing industries usage of water and dyes are very high among the quantity 50% of the dyes are send out to wastewater which causes the environment pollution (Lin et al. 2012). Dyes have global usage in textiles, leather, plastics, paper, food and cosmetic industry to color their products, which could be characterized by strong odor, high chemical oxygen demand (COD) and broad range of pH (Mohabansi et al. 2011). So, the treatment of wastewater generated in textile industry has got great attention because of its considerable impact on creating environment pollution. Textile processing unit effluent is usually released to municipal sewage treatment plants or directly into waterways (Cooper 1993). The conventional techniques currently used can only transplant the non-bio-degradable mass into sludge, which causes other types of pollution which need to be treated further (Stock et al. 2000). In these circumstances, traditional dye wastewater treatment processes have become very demanding for existing plants. As a consequence

A. Tallapaka (✉) · J. Thati · S. Chintha
University College of Technology (A), Osmania University, Hyderabad, India

© Springer Nature Singapore Pte Ltd. 2020
S. K. Ghosh et al. (eds.), *Recent Trends in Waste Water Treatment and Water Resource Management*, https://doi.org/10.1007/978-981-15-0706-9_12

of these problems, advanced oxidation process (AOP) contemplated as an implied technology for dyes' wastewater treatment. The concept of AOP was proposed in the 1980s since then different advanced oxidation technologies have been studied and applied for treatment of industrial wastewater and also for municipal wastewater (Deng and Zhao 2015). Among the available technologies, heterogeneous photocatalysis using an irradiated dissipation of titanium dioxide (TiO_2) is considered as an alternate process for removing organic compounds from wastewater (Chladova et al. 2011).

Oxidative photocatalysis is one of the AOP and an alternative method for collapsing the organic pollutants in air and water. In this process, it includes light, photocatalyst, and an oxidizing agent to achieve the desired chemical reaction (Hassaan and El Nemr 2017). Titanium dioxide is found to be more productive for photocatalytic degradation of organic pollutants due to faster electron transfer to molecular oxygen. TiO_2 is a selective photocatalyst for the environmental engineering application; due to its capacity to provide total mineralization, nontoxic, low-cost, reusable, photostable, alkali, resistance to acid and organic solvents (Ameta et al. 2013; Gültekin and Ince 2007). For TiO_2 preparation, there are many procedures available, often used methods are sol-gel method, hydrothermal processing and gas-phase reaction.

In sol-gel method, high purity materials are used and, when it is transformed into liquid state by mixing with a solvent, it has lot of benefits such as uniformity of products, low cost and easy control of chemical stoichiometry and microstructure (Mohabansi et al. 2011).

Various classes and types of dyes are used in textile industry; among them, Methylene Blue, which is a cationic dye was selected for degradation studies in this work. Because, methylene blue is considered as a toxic dye and it can bring up very harmful influences on the living things due to inhalation. Problems include breathing complications, vomiting, diarrhea, and nausea (Lin et al. 2012). The structural formula of methylene blue is given in Fig. 1 (Mohabansi et al. 2011) (Table 1).

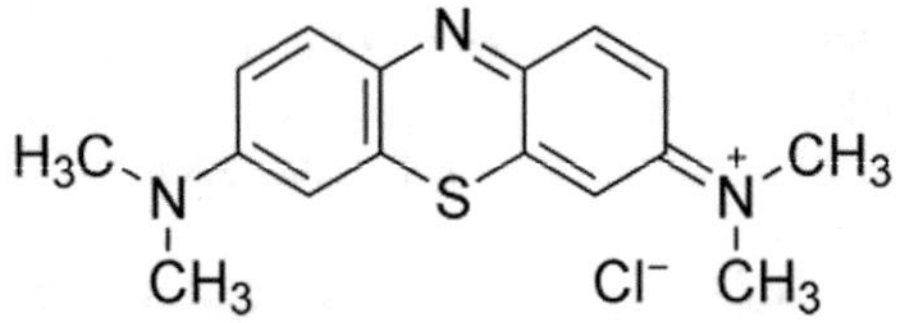

Fig. 1 Methylene blue structure

Table 1 Methylene blue properties (Mohabansi et al. 2011; Deng and Zhao 2015)

Properties of Methylene Blue	
Chemical formula	$C_{16}H_{18}N_3SCl$
Chemical name	3.7-bis(Dimethylamino)-phenothiazin-5-ium chloride
Molecular weight	319.85
Melting point	100–110 °C
Density	1.1 g/cm^3

This current research work mainly aims on the effective removal of MB dye from wastewater using oxidation photocatalysis using titanium dioxide as catalyst.

2 Experimental Work

2.1 Materials

Methylene blue (MB) analytical grade purchased from Fisher scientific, titanium dioxide (TiO_2), and distilled water.

2.2 Method

Experiments were conducted in a borosilicate glass batch reactor of capacity 250 ml placed in a special UV cabinet equipped with UV light emitting 254 nm wavelength. The extent of degradation of MB was analyzed by UV spectrophotometer (Schimadzu model UV-1800). For obtaining calibration data, the MB dye solution concentrations were calibrated initially at 664 nm wavelength corresponding to the maximum absorbance noticed. In the present work, TiO_2 was considered as a photocatalyst for the deterioration of methylene blue dye in wastewater (Lin et al. 2012). Parameters studied were weight of catalyst, concentration of dye, and pH irradiated for 190 min.

- Weight of (TiO_2) nanocatalyst (25, 50, 100 mg).
- Methylene blue concentration of (20, 30, 40 mg/l).
- pH range of (3.0, 7.0, 9.0).

The residual amount of MB was determined by the use of a UV spectrophotometer (Shimadzu, model UV-1800). According to the characteristic absorbance peaks recorded with spectrophotometer, MB concentration was determined at 664 nm by the UV detector (Figs. 2 and 3).

$$\% \text{ degradation} = \frac{C_{MB0} - C_{MB}}{C_{MB0}} \times 100$$

2.3 Preparation of Catalyst

Titanium dioxide (TiO_2) nanocatalyst was produced by the ultrasonic-assisted sol-gel method using titanium (IV) iso-propoxide (98%) as precursor. Briefly, titanium

Fig. 2 Maximum
absorbance of MB dye
solution for different
concentrations at 664 nm

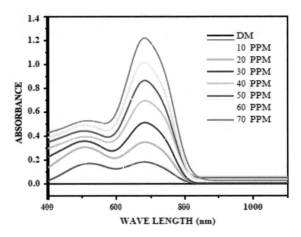

Fig. 3 Calibration curve for
MB dye at 664 nm

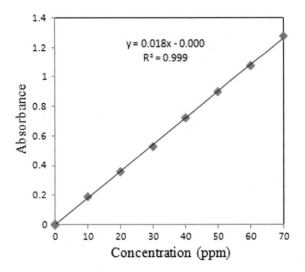

iso-propoxide (TI) is dissolved in 2-Propanol to yield TI solution. A second solution
is then prepared by mixing 2-Propanol with water, and this solution was slowly
adding up to the TI solution to form the TiO$_2$ gel. The gel was changed into a yellow
block crystal after drying at 105 °C for 5 h in oven. The resulting powder was then
subjected to calcination at a temperature of 500 °C for 6 h (Karami 2010).

2.4 Characterization Studies: X-Ray Diffraction (XRD)

The XRD pattern of nanocatalyst (TiO$_2$) in Fig. 4 shows the presence of three main
peaks at $2\theta = 25.3°$ (101), 26.1°, and 37.0°, which is an attributive indication of an

< Group: 12-9-13-Prof-Sailu   Data: TiO2-Ajith >

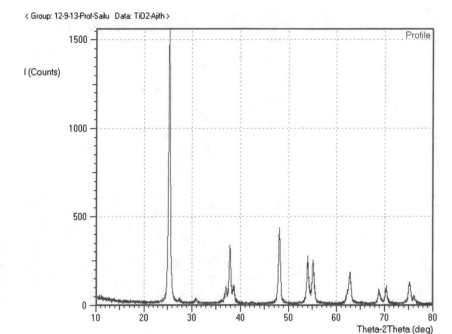

Fig. 4 XRD photograph of (TiO$_2$) catalyst

anatase TiO$_2$. Contrarily, minor diffraction peaks noticeable at $2\theta = 27.5°$ (110) and 31.5°, which correlates to rutile TiO$_2$. The outcome is in correspondence with its mineral configuration bearing anatase to rutile ratio of 83:17 (Karami 2010).

2.5 Analysis by Scanning Electron Microscope (SEM)

Scanning electron microscope is used for the characterization of the morphology of surface and particle size. SEM analysis results the particle size as 156–382 nm (Fig. 5).

Particle size, zeta potential were analyzed by nano particle analyzer (Horiba model SZ-100 series); From the analysis the catalyst (TiO$_2$) mean particle size (388.8 nm) and Zeta potential (-39.6 mV) were noted (Fig. 6).

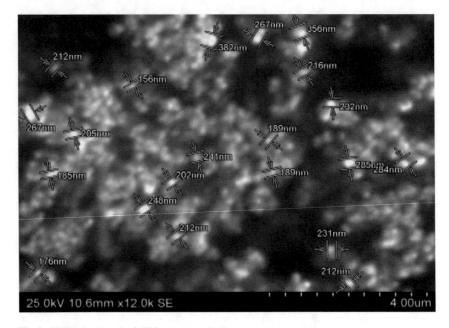

Fig. 5 SEM micrograph of (TiO$_2$) nanocatalyst

3   Results and Discussion

3.1   *Effect of MB Dye Concentrations and TiO$_2$ Catalyst Loading*

The influence of catalyst is shown in Figs. 7 and 8 and it is clear that for nano sized catalyst (TiO$_2$) prepared has substantially removed MB dye with in considerable time of 190 min. A considerable % of removal was observed with 100 mg catalyst for 20 mg/l of MB solution and to the extent of 86%. Using TiO$_2$ as photocatalyst, 20 ppm methylene blue in water was irradiated for 10 h and achieved 57.06–84.14% of degradation (Ameta et al. 2013). This is because of the photocatalytic activity of TiO$_2$ catalyst apparently depends on numerous factors, such as the pore size and number, surface area, the amount of hydroxyl content in TiO$_2$ catalyst, and so on (Yu et al. 2000).

3.2   *Effect of PH*

An important parameter in the photocatalytic reaction is the pH of solution. The effect of pH on the degradation of MB was investigated over a pH range of (3.0, 7.0,

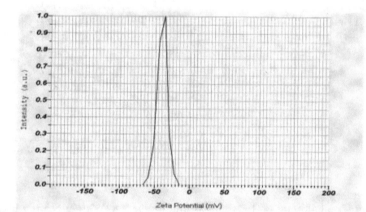

Fig. 6 TiO₂ catalyst nanoparticles analysis

Fig. 7 Methylene blue
degradation with time at
TiO₂ catalyst loading

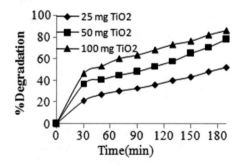

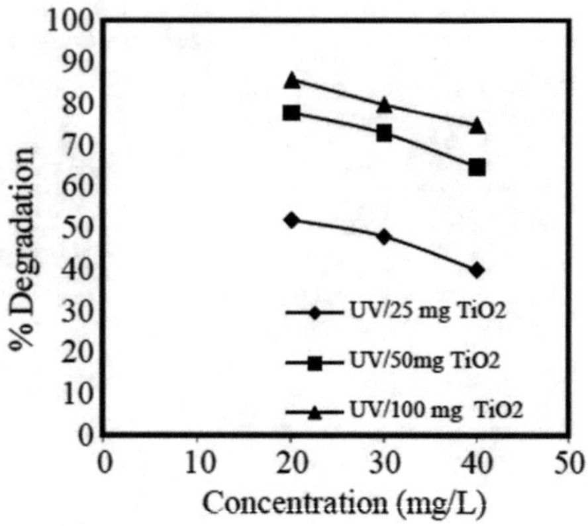

Fig. 8 Methylene blue degradation with concentration at different catalyst loadings

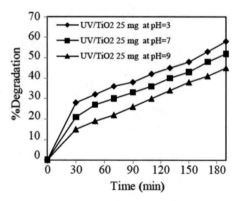

Fig. 9 Methylene blue degradation with time at different pH levels

9.0) using 25 mg/250 ml of TiO_2 in 190 min irradiation. With increasing pH, the degradation of methylene blue has decreased (Fig. 9). Based on nature of the organic pollutants, an increase of the pH will have a positive or negative effect on their degradation rate and consequently the mineralization rate of the solution. The pH is a complicated parameter since it is correlated to the ionization state of the surface and has a significant influence on the removal of organic pollutants (Mohabansi et al. 2011; Aziz et al. 2019).

4 Conclusion

The deterioration of methylene blue dye solution was studied by oxidative photocatalysis. The degradation was strongly influenced by the factors such as TiO_2 catalyst loading, initial dye concentration, and pH of the solution. The maximum degradation (86%) of dye solution (20 mg/L of initial dye concentration) was achieved using

100 mg of TiO_2 in 190 min of irradiation under UV light. The deterioration of methylene blue dye is efficient in acidic solution, and it was more at pH of 3. This method is promising to use for the removal of methylene blue dye present in wastewater.

References

Ameta, A., Ameta, R., & Ahuja, M. (2013). Photocatalytic degradation of methylene blue over ferric Tungstate. *Scientific Reviews & Chemical Communications, 3*(3), 172–180.

Aziz, H. A., Salem, S., & Amr, A. (2019). *Advanced oxidation processes (AOPs) in water and wastewater treatment* (pp. 1–501). IGI Global. (Web. 8 January 2019). https://doi.org/10.4018/978-1-5225-5766-1.

Belgiorno, V., Rizzoa, L., Fatta, D., Della Rocca, C., Lofranoa, G., Nikolaouc, A., & Naddeo, V. (2007). Review on endocrine disrupting-emerging compounds in urban wastewater: Occurrence and removal by photocatalysis and ultrasonic irradiation for wastewater reuse. *Desalination, 215,* 166–176.

Chladova, A, Wiener, J., & Polakova, M. (2011, September 23). Testing the photocatalytic activity of TiO_2 nanoparticles with potassium permanganate solution, Brno, Czech Republic, EU.

Cooper, P. (1993). Removing colour from dyehouse waste waters—A critical review of technology available. *Journal of the Society of Dyers and Colourists, 109,* 97.

Deng, Y., & Zhao, R. (2015). Advanced oxidation processes (AOPs) in wastewater treatment. *Current Pollution Reports, 1*(3), 167–176.

Gültekin, I., & Ince, N. H. (2007). Synthetic endocrine disruptors in the environment and water remediation by advanced oxidation processes. *Journal of Environmental Management, 85,* 816–832.

Hassaan, Mohamed A., & El Nemr, Ahmed. (2017). Advanced oxidation processes for textile wastewater treatment. *International Journal of Photochemistry and Photobiology, 2*(3), 85–93.

Karami, A. (2010). Synthesis of TiO_2 nano powder by the sol-gel method and its use as a photocatalyst. *Journal of the Iranian Chemical Society, 7*(2), S154–S160.

Lin, W. C., Yang, W. D., & Jheng, S. Y. (2012). Photocatalytic degradation of dyes in water using porous nanocrystalline titanium dioxide. *Journal of the Taiwan Institute of Chemical Engineers, 43,* 269–274.

Mohabansi, N. P., Patil, V. B., & Yenkie, N. (2011). A comparative study on photo degradation of methylene blue dye effluent by advanced oxidation process by using TiO_2/ZnO photo catalyst. *Rasayan Journal of Chemistry, 4*(4), 814–819.

Stock, N., Peller, J., Vinodgopal, K., & Kamat, P. V. (2000). Combinative sonolysis and photocatalysis for textile dye degradation. *Environmental Science & Technology, 34,* 1747.

Yu, J., Zhao, X., & Zhao, Q. (2000). Effect of surface structure on photocatalytic activity of TiO_2 thin films prepared by sol–gel method. *Thin Solid Films, 379,* 7–14.

Bioremediation of Textile Azo Dyes by Marine Streptomyces

Bhoodevi Chakravarthi, Vani Mathakala and Uma Maheswari Devi Palempalli

Abstract The present study was carried out to explore the efficiency of marine *Streptomyces sviceus* in the bioremediation of reactive textile azo dyes by considering Congo red-21 as model dye. *Streptomyces sviceus* shows optimum decolourization of Congo red-21 in nutrient glucose broth at 96-h duration under static conditions. Toxicity of the degraded metabolites of Congo red-21 was assessed through phytotoxicity, genotoxicity, acute toxicity and cytotoxicity assays. The *Allium cepa* roots treated with Congo red-21 dye exhibited chromosomal aberrations like polyploidy interphase nucleus and vagrant chromosomes. In contrast, no significant changes were noticed with degraded sample of Congo red-21. These results demonstrate the nontoxic nature of the degraded sample of Congo red-21 on the growth of higher plants and further confirm the bioremediation potential of marine *Streptomyces sviceus*.

Keywords Bioremediation · Mitotic index · Acute toxicity · Streptomyces · MTT assay

1 Introduction

Textile industries predominantly employ chemical dyes compared to natural dyes because of their wide spectrum of colours and cost economy in production due to synthesis. Among the different textile dyes, azo dyes represent 70% and are widely used for dyeing different types of textile fabric material (Chen Cheunbarn et al. 2008). Azo dyes are integral part of textile industry, and they damage ecosystem through textile effluents during dyeing process. The release of azo dyes into the natural environment results into the conversion of azo group to toxic aromatic amines which are highly carcinogenic and mutagenic to humans (Jadhav et al. 2007; Lade et al. 2015). Further, the release of textile dye effluent into water stream changes the pH scale and increases the BOD, COD and total organic carbon of the water bodies (Mathur and Pradeep 2007; Lade et al. 2012).

The conventional wastewater treatment processes are unable to remove the reactive dyes due to their stability, colourfastness and resistance towards degradation.

B. Chakravarthi · V. Mathakala · U. M. D. Palempalli (✉)
Department of Applied Microbiology, Sri Padmavati Mahila Visvavidyalayam, Tirupati, AP, India

© Springer Nature Singapore Pte Ltd. 2020
S. K. Ghosh et al. (eds.), *Recent Trends in Waste Water Treatment and Water Resource Management*, https://doi.org/10.1007/978-981-15-0706-9_13

The recalcitrance of the reactive azo dyes is mainly due to the presence of $(-N = N)$ which are of synthetic origin (Bafana et al. 2008). Due to their synthetic origin and complex fused aromatic structures, the reactive azo dyes are highly difficult to degrade by utilizing conventional effluent, treatment procedures such as pH neutralization, coagulation followed by biological treatment (Anajeneyulu et al. 2005).

For the last ten years, bioremediation is being adopted as an effective alternative method for degradation, mineralization and detoxification of azo dyes. Bioremediation is considered as an eco-friendly and cost competitive alternative to physico-chemical degradation of textile effluents (Raja et al. 2016). The effectiveness of biodegradation depends on the adaptability and bioremediation potential of selected organisms (Gupta et al. 2011). Several microorganism including bacteria, fungi, algae and actinomycetes have been utilized for the degradation and mineralization of azo dyes.

Currently, there has been an intensive research on the application of marine actinomycetes in the degradation and mineralization of azo dyes. Actinomycetes are gram-positive filamentous bacteria widely distributed in terrestrial and marine ecosystem. Actinomycetes have a profound role in the marine system due to their adaptation to the low concentration of carbon substances for their growth and their ability to degrade xenobiotic compounds. The quantitative data on the decolorization and mineralization of reactive azo dyes by actinomycetes and its enzymes is limited. Thus, the development of novel decolorization method by exploiting the marine microbial community is need of the hour for effective removal of azo dyes from textile effluents. Keeping in view the significance of marine actinomycetes, the present study is designed to assess the bioremediation potential of marine *Streptomyces sviceus* by using congo red-21 as a sample dye.

2 Materials and Methods

2.1 Materials

The reactive azo dye Congo red-21 was collected from the retail vendors at Madanapalli and Puttur towns of Chittoor district, A.P., India. *Streptomyces sviceus* was isolated from marine sediment samples collected from the Bay of Bengal coastline of Nellore district of A.P. state. All other chemicals and solvents used were of analytical grade procured from SD fine and Sigma chemicals.

3   Methodolgy

3.1   Decolourization of Congo Red-21 Under Static/Shaking Conditions

Streptomyces sviceus was cultivated for 48 h in 250 ml conical flask containing 100 ml of nutrient glucose broth, and the medium was supplemented with 400 mg/l of Congo red-21. The dye degradation activity of *Streptomyces sviceus* was determined under static and shaking conditions (150 rpm) at 28 $\pm$ 2 °C in orbital shaking incubator. An aliquot (3 ml) of the culture medium was withdrawn at regular time intervals and subjected to centrifugation at 10,000 rpm. The supernatant was collected and used for measuring the percentage of decolorization (Parvez. R et al. 2015).

3.2   Assessing the Toxicity of Dye and Degraded Metabolites

The bioremediation potential of *Streptomyces sviceus* KN_3 was assessed through genotoxicity, acute toxicity and cytotoxicity assay of Congo red dye and its degraded metabolites. Toxicity of the Congo red-21 dye and its degraded metabolites was analysed using *Allium cepa* as in vivo model.

3.3   Cytogenetic Analysis

The cytotoxicity of control and degraded Congo red-21 dye was assessed through the cytological changes in the cells of Allium cepa roots. The seeds were soaked in distilled water for 2 h and further placed on wet filter paper at 23–25 °C for germination under dark conditions. After germination, the Allium cepa seedlings were divided into three groups. Control group treated with distilled water, Group-II with untreated Congo red-21 dye and Group-III with treated Congo red-21 dye. After 72 h of treatment, root tips were carefully cut and fixed in a freshly prepared Carnoy fixative (3:1 v/v absolute alcohol and glacial acetic acid) and stored in 70% ethyl alcohol at 40 °C for 24 h.

Root tips of Allium cepa were thoroughly washed and hydrolyzed in HCl at 60 °C for 4–6 min. The tips were washed and stained with Fulgent reagent for 2 h (Navya et al. 2013). The deeply stained terminals of root tips were squashed in a drop of 45% acetic acid on a glass slide. After covering with cover slip and sealing, the slides were examined for measuring mitotic indices and frequency of mitotic abnormalities.

$$\text{Mitotic index(MI)} = \frac{(\text{Number of dividing cells in all phases of mitosis}) \times 100}{\text{Total number of counted cells}}$$

$$\text{Abnormalities of Mitosis (\%)} = \frac{(\text{Total number of abnormal cells}) \times 100}{\text{Total number of dividing cells}}$$

3.4 Phytotoxicity

Toxicity of Congo red-21 and its degradation metabolites on germination of plants was assessed by using green gram and fenugreek seeds. The seeds were treated daily with 5 ml of parent Congo red-21 dye (400 mg/L) and its degraded metabolites. The length of root, shoot and the percentage of seed germination were recorded after 10 days.

3.5 Acute Toxicity Test

To study the effect of Congo red dye and its degradation metabolites, on fish mortality and their behaviour, acute toxicity test was conducted using zebrafish as in vivo model. The Congo re-21 dye degraded by the *Streptomyces sviceus* strain KN$_3$ was subjected to centrifugation at 10,000 rpm for 30 min. The supernatant was collected, filtered through membrane filters. About 100 ml of clear filtrate was taken in a beaker, and then, 24-h-old neonates of zebrafish were added. The experiment was carried out at room temperature for 48 h under dark conditions, and the number of immobile organisms was counted after exposing to light for 30 s.

3.6 Cytotoxic Activity Assay

The cytotoxic activity of the degraded metabolites was assessed by MTT assay (Lavanya et al. 2014) The effect of degraded dye sample on the cell viability was assessed through mitochondrial dehydrogenase activity (3-(4, 5-dimethylthiazol-2-yl)-2, 5-diphenyltetrazolium bromide) (Claiborne et al. 1979). Raw 264.7 cells (**Murine macrophages**) were seeded (2×10^4 cells/well) and incubated for 24 h at 37 °C in 5% CO_2 atmosphere for the adherence of the cells to the flat bottom of the 96-well plates. The cells were treated with different concentrations of Congo red-21 control and degraded sample (25, 50, 75 and 100 µg/ml) for 24 h. At the end of the incubation, media was removed and washed with 0.01 M PBS (pH 7.4) followed by the addition of 25 µl of MTT solution (5 mg/ml in PBS) to each well and incubated for 3 h at 37 °C. After incubation 0.1 ml of extraction buffer (20% SDS in 50% dimethylformamide) was added and incubated for 30 min to extract the purple-blue MTT Formosan crystals. The absorbance was measured at 540 nm using micro-plate reader (Biotek-255907), and the percentage of cell survival was calculated.

4 Results

The effect of static and shaking conditions on rate of decolorization of reactive azo dyes was analysed at different time intervals (0–96 h). The maximum decolorization of Congo red-21 was noticed at all-time points under static conditions. *Streptomyces sviceus* showed 70% of decolorization under shaking condition, and the rate of decolourization was accelerated to 90% under static (Fig. 1).

4.1 *Effect of Congo Red-21 and Its Degraded Metabolites on Root Elongation*

The effect of Congo red-21 control dye and its degraded sample was tested on the root elongation of Allium cepa. The root growth was totally blocked in the presence of control dye, and on the other hand, the effect of dye properties on the root growth was reduced to 50% after degradation by *Streptomyces sviceus* (Fig. 2). The average length of Allium cepa root accounts to 4.36 cm after 72 h of growth in the presence of water. In contrast, the dye control restricted the root growth to a level of 0.85 cm, whereas in the presence of degraded sample, the root growth was 2.86 cm which was found to be nearly 60% of root length with water (Table 1).

4.2 *Cytogenetic Analysis of Congo Red-21 Dye and Its Degraded Products*

Allium cepa is considered as a remarkable genetic model to assess the genotoxic effects of Congo red-21 and its degraded metabolites. The inhibitory effect of Congo red-21 dye on elongation of root length was due to the aberration in mitosis of growing cells. From the microscopic observation of the section cuttings, it was observed that the number of undivided cells was high in Congo red dye treated root tips. The

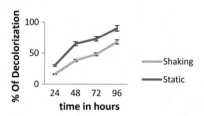

Fig. 1 Rate of decolorization of Congo red-21 under static and shaking conditions [Values are expressed in the mean ± SEM of three independent experiments. Statistical significance 1% level ($p \leq 0.01$)]

Fig. 2 Influence of Congo red-21 dye and its metabolites on root elongation of *Allium cepa*. **a** Water control, **b** Congo red-21 dye control, **c** Congo red-21 dye degraded sample

Table 1 Elongation of Allium cepa roots

Particulars		Root length $\pm$ SD (cm)
Water control		4.36 ± 0.26
CR-01 dye control		0.85 ± 0.32
CR-01 degraded sample	(400 mg/l)	2.86 ± 0.30

number of dividing cells and mitotic index (MI) was found to be high in the roots treated with degraded dye sample which indicates the nontoxic nature of degraded dye was represented in Table 2.

In order to detect the genotoxicity of Congo red-21 and its metabolites, the four phases of cell division were considered to measure the chromosomal aberration. The Allium cepa roots treated with Congo red-21 dye exhibited polyploidy interphase nucleus, C-metaphase, spindle disturbance at anaphase, disoriented chromosome, sticky chromosome and vagrant chromosomes (Fig. 3). In contrast, no significant changes were noticed with degraded sample of Congo red-21. The results demonstrate the nontoxic nature of the degraded sample on the growth of higher plants and further confirm the bioremediation potential of *Streptomyces sviceus* (Fig. 4).

Table 2 Mitotic index of Allium cepa root tip cells exposed to Congo red-21 dye and its degraded metabolites

Treatment	Number of cells analysed	Number of dividing cells	% of Mitotic index
Water	172	82	47.67
Congo red-21 control dye (400 mg/l)	91	9	9.89
Congo red-21 degraded with *Streptomyces sviceus*	158	45	28.48

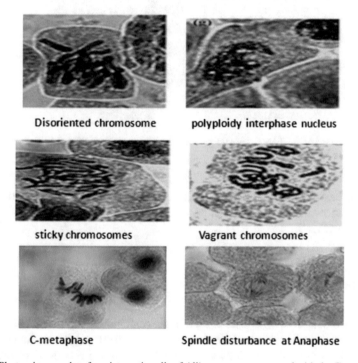

Disoriented chromosome   polyploidy interphase nucleus

sticky chromosomes   Vagrant chromosomes

C-metaphase   Spindle disturbance at Anaphase

Fig. 3 Photomicrographs of meristematic cells of *Allium cepa* roots treated with the Congo red-21 dye

4.3 Phytotoxicity of Congo Red-21 Dye and Its Degraded Products

The germination percentage of green gram seeds was studied to assess the phytotoxicity of degraded metabolites of Congo red-21. The seedlings of green gram showed 2.1-fold (80%) higher germination rates with degraded Congo red-21 sample in comparison with control dye (38%) (Table 3). The length of plumule and radi-

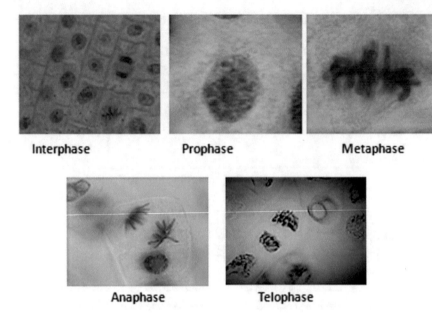

Interphase   Prophase   Metaphase

Anaphase   Telophase

Fig. 4 Photomicrophs of meristematic cells of Allium cepa roots treated with Congo red-21 degraded with *Streptomyces sviceus*

cle was observed to be significantly affected by Congo red-21 parent dye than their metabolites, indicating less toxic nature of degradation metabolites as compared with parent dyes (Fig. 5).

4.4 Acute Toxicity Test

The acute toxicity assay was carried out in water containing Congo red-21 (400 mg/L) and different concentrations (100, 250 and 400 mg/l) of degraded Congo red-21 sample. The results of the toxicity assay are tabulated as per cent death of zebrafish. The percentage mortality was found to be high at all concentrations. In contrast, the degraded sample with 400 mg/l dye concentration showed mortality from 3 to 11% at different time intervals (Table 4). These results suggest that release of untreated Congo red-21 dye is a critical concern to the environment and human health, whereas dye treatment with *Streptomyces sviceus* could be used for detoxification and mineralization of azo dyes (Fig. 6).

Table 3 Influence of Congo red-21 dye and its degraded metabolites on germination of green gram seeds

Samples	Percentage of germination										Average shoot length (cm)	Average root length (cm)
	Day1	Day2	Day3	Day4	Day5	Day6	Day7	Day8	Day9	Day10		
Control (water)	0	29	42	53	67	79	88	100	100	100	17.9 ± 0.351	3.5 ± 0.351
Degraded Congo red-21 Dye	0	20	31	39	47	54	62	69	77	80	16.1 ± 0.833	2.9 ± 0.436
Congo red dye	0	6	10	13	18	24	30	32	36	38	10.2 ± 0.100	1.8 ± 0.361

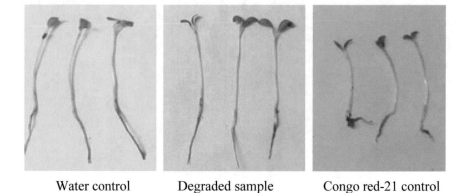

Water control   Degraded sample   Congo red-21 control

Fig. 5 Effect of Congo red-21 dye and its metabolites on the growth of green gram seedlings

Table 4 Effect of Congo red-21 dye and its degraded metabolites on mortality of fish

Time (h)	Mortality degraded sample (%)				
	Distilled water	Congo red-21 control 400 mg/l	Degraded sample		
			100 mg/l	200 mg/l	400 mg/l
6	0	52 ± 0.2	0	0	9 ± 0.1
12	0	80 ± 0.4	0	10 ± 0.4	17 ± 0.1
24	0	87 ± 0.3	0	22 ± 0.6	40 ± 0.6

Congo red-21 dye (400mg/L)   Degraded sample (400mg/L)

Fig. 6 Toxicity of Congo red-21 dye and its degraded metabolites on existence of zebrafish

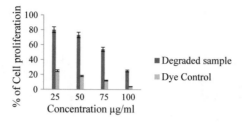

Fig. 7 Cytotoxicity of Congo red-21 and its decolorized sample against RAW 264.7 cells (values are expressed as the mean ± SEM of three independent experiments statistical significance 1% level ($p \leq 0.01$))

4.5 Cytotoxicity of Degraded Products

MTT assay was carried out to analyse the toxic nature of Congo red-21 dye and degraded metabolites on the cell proliferation of RAW 264.7 macrophages. The assay was performed with different concentrations (25–100 µg/l) of degraded product.

The proliferation of RAW 264.7 macrophage cell lines was inhibited due to the toxic nature of Congo red-21 control dye. In contrary to the parent dye, 30–80% of cell proliferation was noticed with varying concentration of degraded samples shown in (Fig. 7). The visibility of cell proliferation even at higher concentration of degraded products confirms mild interference and nontoxic nature of the degraded product and further, the bioremediation potential of *Streptomyces sviceus* strain KN$_3$ (Fig. 8).

5 Discussion

The discharge of azo dyes, through textile effluents, into the water stream and soil may escalate the pollution. The environmental and health hazards of textile dye effluents have been a subject to research scrutiny for a long time. The present research originated based on the need to gaze into bioremediation potential of *Streptomyces sviceus isolated from marine sediment samples.*

The cytotoxicity, mutagenicity of the azo dyes were analysed though root shape, elongation, mitotic index and chromosomal aberrations (Misra and Fridovich 1972). The decrease in root growth to an extent of 45% represents the toxic of the xenobiotic compounds on the plants (Wierzbicka M. 1999). Our results demonstrate the cytotoxic nature of Congo red-21 dye and nontoxic properties of its degraded metabolites. Mitotic index (MI) is the hallmark for estimating the frequency of cellular division and inhibition of mitotic activities which are the indices to mark the cytotoxicity of the synthetic as well as natural compounds (Amin et al. 2002). Reduced MI is due to the inhibition of DNA synthesis, and decrease in mitotic index is considered as an important indicator to pollution in affected environments, particularly at the sites

Congo red-21 Dye

Congo red-21 degraded with
Streptomyces sviceus

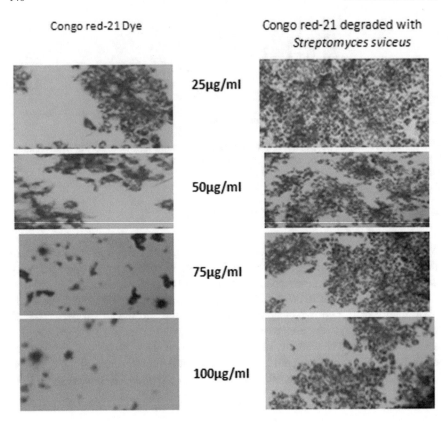

25µg/ml

50µg/ml

75µg/ml

100µg/ml

Fig. 8 Proliferation of RAW 264.7 cells in the presence of Congo red-21 dye and its degraded products

contaminated with azo dyes (Marcano et al. 2004). In the experimental study, mitotic index of degraded Congo red-21 sample was lower than that of water control, and in contrast, the mitotic index was improved significantly in *Allium cepa* meristematic cells treated with degraded dye sample. Decrease in mitotic index was reported with orange red dye and CI disperses blue 291 (Tsuboy et al. 2007). The reduced MI is attributed to the onset of prophase, blocking one or more mitotic phases (Pillai et al. 2011). Chromosomal aberrations were assessed to detect the genotoxic activity of Congo red-21 dye and its degraded metabolites. The meristematic cells treated with Congo red-21 dye represented the most common aberration like polyploidy interphase nucleus, C-metaphase, sticky metaphase with chromosomal adherence, spindle disturbance at anaphase and vagrant chromosome. The chromosome adherence may be an indicator for genotoxic effect of azo dye which may result in irreversible cell damage and necrosis. The disturbance at spindle prevents the migration of the chromosome towards the pole and chromosomes become attached to results to each other through bridges and results in the condensation and chromosome, stickiness (Pillai et al. 2011). Chromosome adherence or stickiness indicates the toxic effects of xeno-biotic compounds which are irreversible and leading to death (Kalyani et al. 2008). This report is in correlation with our results, where reduction in mitotic index was noticed with Congo red-21 dye. These results clearly suggest the cytotoxic effect of Congo red-21 dye and the nontoxic nature of degraded metabolites.

The bioremediation potential of *Streptomyces sviceus* was assessed through phy-totoxicity assay. The phytotoxicity of the dye was detected by measuring the potential of Congo red-21 dye and its metabolites on the germination of green gram seeds. The results show that the degraded metabolites were not toxic to germination of green gram seedlings and good growth of plants was observed when compared to control.

Evaluating the toxicity of azo dyes before and after degradation of bacteria is necessary as the dye and their metabolities are phytotoxic and toxic to aquatic and human health (Phugare et al. 2011). Acute toxicity tests by using Daphnia magna or zebrafish as in vivo model are well-known screening method for analysing the lethal toxicity of xenobiotics and their metabolites to animal and human health. The relative toxicity of Congo red-21 and its degradation product after treatment with *Streptomyces sviceus* strain KN_3 were studied by acute toxicity with zebrafish.

The cytotoxic effect of Congo red-21 dye and its degraded metabolites on RAW 264.7 macrophages was studied by MTT assay. Befana and chakrabarti 2008 reported the cytotoxicity of DR 28 dyes at different levels of biodegradation by biodegradation by *Bacillus velezensis*.

The mouse fibroblast cell line was used as in vitro model for assessing the cytotox-icity of red 141 and reactive red 2 dye degraded by *Bacillus lentus* (Kumar and Bhat 2012). Our results explained the nontoxic nature of degraded sample by promoting cell proliferation of RAW 264.7 cells.

6   Conclusion

From the present study, it can be concluded that *Streptomyces sviceus* is highly potential in the detoxification of the reactive azo dyes, particularly Congo red-21 dye. Thus, the present research study provides ample evidence for bioremediation of azo dyes by marine Actinobacteria.

References

Amin, A. W. (2002). Cytotoxicity testing of sewage water treatment using Allium cepa chromosome aberration assay. *Pakistan Journal of Biological Sciences, 5,* 184–188.

Anjaneyulu, Y., Chary, N. S., & Raj, D. S. S. (2005). Decolorization of industrial effluents—available methods and emerging technologies-a riview. *4,* 245–273.

Bafana, A., Krishnamurthi, K., Devi, S., & Chakrabarti, T. (2008). Biological decolourization of C.I. directs black 38 by *E. gallinarum. Journal of Hazard Materials, 157,* 183–193.

Chen Cheunbarn, T., Cheunbarn, S., & Khumjai, T. (2008). Prospects of bacterial granule for treatment of real textile industrial waste water. *International Journal of Agricultural Biology, 10,* 689–692.

Claiborne, A., Malinowski, D. P., & Fridovich, I. (1979). Purification and characterization of hydroperoxidase II of Escherichia coli B. *Journal of Biological Chemistry, 254,* 11664–11668.

Gupta, N., Tripathi, A. K., & Harsh, N. S. K. (2011). Bioremediation of cottontextile effluent using fungi. *Bulletin of Environment, Pharmacology and Life Sciences, 1*(15), 19.

Jadhav, J. P., Parshetti, G. K., Kalme, S. D., & Govindwar, S. P. (2007). Decolourization of azo dye methyl red by *Saccharomyces cerevisiae* MTCC463. *Chemosphere, 68,* 394–400.

Kalyani, D. C., Telke, A. A., Dhanve, R. S., & Jadhav, J. P. (2008). Ecofriendly biodegradation and detoxification of reactive red 2 textile dye by newly isolated *Pseudomonas sp.* SUK1. *Journal of Hazardous Materials, 163,* 735.

Kumar, P. G. N., & Bhat, S. K. (2012). Fungal degradation of azo dye- Red 3BN and optimization of physico-chemical parameters, ISCA. *Journal of biological sciences, 1,* 17–24.

Lade, H., Kadam, A., Paul, D., & Govindwar, S. (2015). Biodegradation and detoxification of textile azo dyes by bacterial consortium under sequential icroaerophilic/aerobic processes. *ECLI Journal, 14,* 158–174.

Lade, H. S., Waghmode, T. R., Kadam, A. A., & Govindwar, S. P. (2012). Enhanced biodegradation and detoxification of disperse azo dye Rubine GFL and textile industry effluent by defined fungal-bacterial consortium. *International Biodeterioration and Biodegradation, 72,* 94–107.

Lavanya, C., Rajesh, D., Sunil, C., & Sarita, S. (2014). Degradation of toxic dyes: A review. *International Journal of Current Microbiology Applied Science, 3*(6), 189–199.

Marcano, L., Carruyo, I., Del Campo, A., & Montiel, X. (2004). Cytotoxicity and mode of action of maleic hydrazide in root tips of Allium cepa L. *Environmental Research, 94,* 221–226.

Mathur, N., & Bhatnagar, P. (2007). Mutagenicity assessment of textile dyes from Sanganer (Rajasthan). *Journal of Environmental Biology, 28,* 123–126.

Misra, H. P., & Fridovich, I. (1972). The role of superoxide anion in the autoxidation of epinephrine and a sample assay for superoxide dismutase. *Journal of Biological Chemistry, 247,* 3170–3175.

Navya, A., Prasad, H., Rashmi, H., Uma Maheswari Devi, P. (2013). Antiproliferation Of Pro-Inflammatory Genes By α-mangostin on MDA-MB-231 human Breast cancer cell line. *IJAPR, 4,* 2222–9.

Parvez, R., & Uma Maheswari Devi, P. (2015). Decolorization and detoxification of reactive azo dyes by *Saccharothrix aerocolonigenes* TE5. *Journal of Applied Environmental Microbiology, 3*(2), 58–62.

Pillai, J. S., Danesh, N., Puttaiah, E. T, & Girish, K. (2011). Microbial diversity in solid waste molasses of sugar industry, Aranthangi, Tamilnadu. *International Journal of Environmental Science, 2*(2), 723–730.

Phugare, S. S., Kalyani, D. C., Patil, A. V., Jadhav, J. P. (2011). Textile dye degradation by bacterial consortium and subsequent toxicological analysis of dye and dye metabolites using cytotoxicity, genotoxicity and oxidative stress studies. *Journal of Hazardous Materials, 186*(1), 713–723.

Raja, M. M., Raja, A., Salique, S. M., & Gajalakshmi P. (2016). Studies on effect of marine actinomycetes on amido black (azo dye) decolorization, *Journal of Chemical and Pharmaceutical Research, 8*(8), 640–644.

Tsuboy, M. S., Anjeli, J. P. F., Mantovani, M. S., Knasmiiller, S., Umbuzeiro, G. A., & Ribeiro, L. R. (2007). Genotoxic, mutagenic and cytotoxic effects of the commercial dye CI Disperse Blue 291 in the human hepatic cell line HepG2. *Toxicology in Vitro, 21,* 1650–1655.

Wierzbicka, M. (1999). The effect of lead on the cell cycle in the root meristem of Allium cepa L. *Protoplasma, 207*(3–4), 186–194.

Use of Sewage to Restore Man-Made Waterbodies—Nutrient and Energy Flow Regulation Approaches to Enabling Sustainability

Seema Sukhani and Hoysall Chanakya

Abstract Sewage ingress into lakes and man-made waterbodies have for long been considered a "Tragedy of the Commons", and numerous attempts have been made to avoid them. Agriculture development and urbanization send nutrient and C laden run-off/sewage into these waterbodies causing eutrophication. This study shows that with minor interventions involving anaerobic digestion for C/N removal and managing trophic levels to lower bacterial biomass at the inlet points can reduce the daily nutrient and energy budgets into these waterbodies while providing much needed water to compensate for daily evaporation (2.5–4 m/year) and maintaining water levels in them. From an understanding of regulating trophic levels, this minimal nutrient and energy input to the waterbody can be managed to raise fish and birds or only harvested fish so as to primarily keep the water quality at a desirable level while also sustaining the waterbody round the year without need for freshwater inputs for long periods, thus converting the "tragedy of the commons" to a "blessing of the commons".

Keywords Man-made waterbodies · Restoration · Lakes · Anaerobic digestion · Protozoa · Ecosystem · Trophic levels · International society of waste management · Air and water

1 Introduction

Sewage ingress or its increasing use to fill up man-made waterbodies alters the balance of the aquatic or near freshwater ecosystems such as lakes and tanks causing them to function less efficiently. It contains significant quantities of nutrients which are otherwise limiting in natural waterbodies and mechanisms to handle these needs to be sought. Such nutrient ingress results in higher primary productivity in terms of excess growth of algae, plants and cyanobacteria and this phenomenon is known as eutrophication. Disposal of sewage has been a major issue in urban and peri-urban areas and its entry into waterbodies is becoming inevitable. Nutrient inputs into such

S. Sukhani · H. Chanakya (✉)
Indian Institute of Science, Bangalore, India
e-mail: chanakya@iisc.ac.in

© Springer Nature Singapore Pte Ltd. 2020
S. K. Ghosh et al. (eds.), *Recent Trends in Waste Water Treatment and Water Resource Management*, https://doi.org/10.1007/978-981-15-0706-9_14

waterbodies have always been considered a liability, while in reality it can now be converted into a blessing. With installation of self-managing sewage treatment systems within a lake, most of these nutrients can be converted into useable resources in terms of energy, fodder and human food including food for other animals (biogas, algal biomass, fish and water birds). However, there is a gap in understanding the balance between nutrient/C loads in effluent of treatment system and their effect on aquatic ecology and self-purification capacity of waterbodies. In this paper, we attempt to show how to convert this "Hardins-Tragedy of the Commons" to a "Blessing of the Commons".

It is now well understood that aquatic ecosystems such as lakes and tanks operate as semi-closed input-output ecosystems (Hannon 1973). The focus while understanding management of lakes had been directed to eutrophication (primary producers), predators (zooplanktons) and improving water quality. Effects of primary producers both in bottom-up (nutrient input to fish/birds) and top-down cascade (fishes-phytoplankton/bacteria) types of aquatic systems have been used in management of tanks. Not until recently, new water quality indicators and control methods had been investigated, and one of emerging approach now uses fish biomass and its harvesting as an important indicator for nutrient harvest and control systems for water quality management (Sosa-López et al. 2005). Such indicators and knowledge about different trophic cascades can be a decisive factor for site-specific management plans for solving water quality problems and re-establishing biotic community (Hansson et al. 1998). However, very little research effort is invested in understanding changes among food web properties across environmental conditions of such man-made aquatic ecosystems. Inspiration can be taken from marine biology where focus has not only been on establishing direct control on producers but integration of other trophic levels and developing a holistic picture. Thus, the completion of a food chain provides an equally effective control mechanism for reducing eutrophication or nutrient ingress in such aquatic systems. Another important factor which has not been receiving proper attention is the impact of secondary treated water on food webs. Most of the organisms' energy is stored in organic carbon bonds within the plant and animal biomass and ~45–55% of dry weight of living entities in freshwater accounts for carbon (Vitousek et al. 1997). This C can very well said to be currency of energy transfer (Dodds 2002). Thus, here we will attempt to understand carbon balance and metabolism at different trophic levels of lake which have an integrated two-stage input sewage treatment unit upstream and trophic levels as shown in Fig. 1.

The main objective is to understand change in flow of carbon and organisms from grazing or production food chain to mineral food chain when selective biological intervention is made and how subsequent trophic levels will be affected.

2 Discussion

2.1 Raw sewage cannot be let directly into the lakes. In Fig. 1, two scenarios have been shown which highlight the effect of quality of influent on lake ecology.

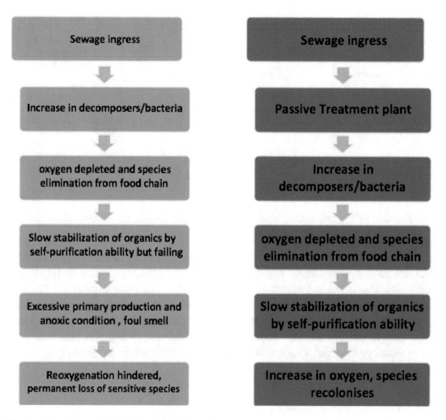

Fig. 1 Effect of quality of influent on lake ecology

1. Near saturation or more load will cause DO to drop significantly. The self-purification process is slow (as the only basis of material exchange is through diffusion and organisms), thus a considerable area from point of discharge will seem polluted. Often the only organisms to flourish will be sewage bacteria, fungus, certain worms and fly larvae. Some species of animals (catfish) and plants (floating hydrophytes, e.g. water hyacinth) may also survive, but it will eradicate endemic species in that area especially fishes. Facultative anaerobic bacteria also thrive in such environment giving off foul smell. Phytoplankton, which can survive in severe organic pollution, will start metabolizing intermediates/exudates (acetate) of anaerobes, hindering reoxygenation. Without any aid and continuous discharge, lake area will remain anoxic and obviously polluted.

2. Treatment plant near the discharge point will bring the influent quality near sub-saturation level. Any leftover organic pollutant apparent after this point will disappear within a short distance in the lake through natural process. Mild pollution can be actually beneficial to waterbody, as it will increase the nutrient and energy supply to micro-organisms and enhance primary productivity. It will not only make the water reusable, but also give returns in terms of energy-rich algal

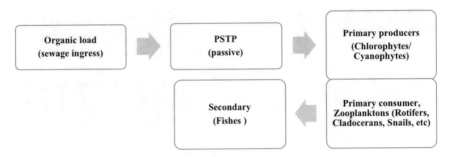

Fig. 2 Schematic of treatment plant and trophic levels in lake

biomass or animal fodder and top most trophic level, i.e. fishes which can be used for human/animal consumption.

2.2 We are using trophic cascade concept to understand aquatic ecology modification due to changes in water source and using simple 3-level cascade in lakes after treatment. This cascade from phytoplankton to fish attempts to simplify the true representation of real-world complexity of food web for the current discussion without diminishing the importance to look at this as a complex interlinked system. Thus, these generalizations will be a foundation in understanding complex interactions of other minor groups and their interactors with the ones mentioned above (Fig. 2).

2.1 Passive Treatment Sewage Plant

Proposed unit is two stages with first stage to be anaerobic and second stage to be phytoflagellate/protozoan driven. In the anaerobic reactor, the phenomenon causing mixing is sludge upwelling as shown in Fig. 3. The details of design and laboratory scale systems that provide the necessary mass/energy balance are published elsewhere.

This reaction system is followed by action of phytoflagellates and protozoans, which play a dual role being grazers on planktons and heterotrophic bacteria, while also stabilizing inorganic carbon through photosynthesis. These communities may itself regulate plankton biomass and turbidity by grazing. Being flagellated, they have the capability of vertical and horizontal movements, which in turn makes them ideal bio-convectors. However, they also pose a problem of an inability to generalize the process with regards to compartmentalizing nutrients and energy to each trophic level. Nevertheless, they are important actors in treatment of sewage and their functional role in a full-scale system is another study objective which will be in focus for future studies.

As mentioned in an earlier publication, the peak wastewater input required to maintain a 10 ha waterbody would be 1 MLD that compensates for the evaporation from the waterbody surface. Schematic of reduction in carbon and other nutrients

Fig. 3 Sludge upwelling in anaerobic reactor increases mixing with fresh input water, increases active bacterial number over a hundred-fold and reduces holding time to 8–16 h

is shown in Fig. 4. The reduction in nutrients has been extrapolated from various lab and field studies (plant under construction). The dominant phytoflagellates are mostly Euglenoids.

Total residence of water in treatment unit is 3 days, in which approximately 75% carbon, 50% nitrogen and 25% phosphorus is reduced. Partially treated water from this point enters the lake, where total residence is 263 d. Most likely because of light availability and soluble nutrients, Chlorophyceae will follow. Nitrogen and phosphorus content is about 10–17 and 1–1.5% of algal DW (Wang et al. 2010). Thus, considering 50% nutrient transfer, per day primary production is expected to be ~30–150 kg/d, which is well under the grazing limit by primary consumer.

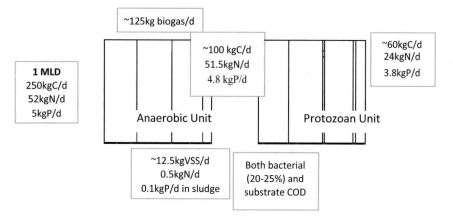

Fig. 4 Overall schematic of PSTP and trophic levels

2.4 Rotifers, cladocerans, snails and crustaceans are major groups of primary consumer, but cladocera are key players in lakes/waterbodies and dominate the zooplankton community both in species, biomass and abundance (Walseng et al. 2006). With their large body sizes, edibility for planktivorous fishes and faster clearance rates, Daphnia spp. (cladocerans) are major actors in trophic cascade (McQueen et al. 1986). Large Daphnia have been reported to filter efficiently algae with cell diameters between 1 and 30 μm and are a major grazer. For this reason, in this paper, Daphnia are used as representative of all grazers and primary consumers. They have a capability to consume ~0.18 mg algae/day per cell (calculated from Mcmahon and Rigler 1965), and their natural abundance is in range of 106–7 cells/ml. Considering a high value of 50% carbon transfer, biomass of grazer is expected to be 30–70 kg/d. This in turn translates to fish productivity of ~15–30 kg/day. Thus, total revenue which can be generated from a 1 MLD plant is 125 kg biogas (60% methane), 100 kg algae dry weight and ~20 kg fish. As for bacteria, viruses and saprophytes, current knowledge about their diversity is still limited, and hence, they have not been addressed here, although clearly they are key components of food web. Fish production can either be controlled by appropriate harvesting techniques for human use or the waterbody may be allowed for bird preserve with an appropriate stocking density of different feeding types. This latter approach not only avoids human consumption but can increase avian biodiversity in the location. This waterbody when used for education or recreation can provide higher daily revenue than fish harvest and can be made a lot more economically sustainable and has been discussed previously.

3   Conclusion

Understanding of phytoflagellates in slightly intervened environment and their trophic role and interaction with other communities is still an open area of research. This, when combined with an anaerobic reactor to remove organic C and some N, can be effectively used to restore and sustainably maintain waterbodies even in areas where sewage/nutrient ingress into aquatic systems threaten waterbody health and sustainability. Man-made lakes and tanks receiving sewage (treated or raw) possess a status far from freshwater status and can be sustained with this approach wherein, in future some foundational works need to be done. Real-time understanding of all trophic levels under various nutrient conditions and changes with respect to abiotic factors has yet to be gained. There is thus a need to modify our current understanding and approach to use of water quality indicators and traditional trophic levels established for freshwater for these man-made waterbodies.

References

Dodds, W. K. (2002). *Freshwater ecology: Concepts and environmental Applications*. Academic Press.

Hannon, B. (1973). The structure of ecosystems. *Journal of Theoretical Biology, 41*(3), 535–546. https://doi.org/10.1016/0022-5193(73)90060-X.

Hansson, L.-A., Annadotter, H., Bergman, E., Hamrin, S. F., Jeppesen, E., Kairesalo, T., & Strand, J. (1998). Minireview: Biomanipulation as an application of food-chain theory: Constraints, synthesis, and recommendations for temperate lakes. *Ecosystems, 1*(6), 558–574. https://doi.org/10.1007/S100219900051.

Mcmahon, J. W., & Rigler, F. H. (1965). Feeding rate of daphnia magna straus in different foods labeled with radioactive phosphorus 1. *Limnology and Oceanography, 10*(1), 105–113. https://doi.org/10.4319/Lo.1965.10.1.0105.

Mcqueen, D. J., Post, J. R., & Mills, E. L. (1986). Trophic relationships in freshwater pelagic ecosystems. *Canadian Journal of Fisheries and Aquatic Sciences, 43*(8), 1581. https://doi.org/10.1139/F86-195.

Sosa-López, A., Mouillot, D., Do Chi, T., & Ramos-Miranda, J. (2005). Ecological indicators based on fish biomass distribution along trophic levels: An application to the Terminos coastal lagoon, Mexico. *ICES Journal of Marine Science, 62*(3), 453–458. https://doi.org/10.1016/j.icesjms.2004.12.004.

Vitousek, P. M., Aber, J. D., Howarth, R. W., Likens, G. E., Matson, P. A., Schindler, D. W., & Tilman, D. G. (1997). Human alteration of the global nitrogen cycle: sources and consequences. *Ecological Applications, 7*(3), 737–750. https://doi.org/10.1890/1051-0761(1997)007%5b0737:Haotgn%5d2.0.Co;2.

Walseng, B., Hessen, D. O., Halvorsen, G., & Schartau, A. K. (2006). Major contribution from littoral crustaceans to zooplankton species richness in lakes. *Limnology and Oceanography, 51*(6), 2600–2606. https://doi.org/10.4319/Lo.2006.51.6.2600.

Wang, L., Min, M., Li, Y., Chen, P., Chen, Y., Liu, Y., Wang, Y., & Ruan, R. (2010). Cultivation of green algae chlorella sp. In different wastewaters from municipal wastewater treatment plant. *Applied Biochemistry and Biotechnology, 162*(4), 1174–1186. https://doi.org/10.1007/S12010-009-8866-7.

Two-Stage Passive High-Throughput Sustainable Sewage Treatment Process: Laboratory Study and Future Scope

Seema Sukhani and Hoysall Chanakya

Abstract Large volumes of sewage need to be handled in the emerging SDG 6 scenario, and low energy/maintenance options are required for distributed and unattended situations. In this direction, a novel two-stage system was developed consisting of a primary anaerobic digestion followed by protozoan/phytoflagellate bacteria/residual OM (organic material) removal stage. Anaerobic purification by sludge upwelling found naturally in idle lagoons was firstly emulated and hastened by encouraging deeper sludge beds that upwell, disperse sludge and maintain high MLVSS without need for mechanical sludge recirculation. In the second stage, a Euglena-driven system that removes mixotrophically, bacteria and residual solids provide reasonably acceptable treated water quality at a HRT < 2d. In this paper, we discuss the preliminary findings. Overall reduction of 70–80% COD, 55% TKN, 25% P and >90% turbidity was achieved. While optimization and in-depth understanding of functionality are being investigated, this system appears to be a good alternative to conventional treatment technologies where space constraints are less and distributed treatment with low interventions is required.

Keywords Wastewater treatment · Passive natural process · Protozoa and phytoflagellates · Anaerobic treatment

1 Introduction

As the world seeks to address sustainable development goals (SDG6) such as sanitation and clean water, providing a household level or decentralized village/town level sewage treatment systems become very important. Sustainability criteria demand the use of less expensive systems as well as low energy options. Towards this, passive decentralized systems are essential where treated water has qualities acceptable at least for surface irrigation or local reuse. The problem is exacerbated when the barely treated sewage flows into nearby natural and man-made water bodies such as lakes and tanks which are generally deprived of natural run-off that was prevalent in the

S. Sukhani · H. Chanakya (✉)
Indian Institute of Science, Bangalore, India
e-mail: chanakya@iisc.ac.in

© Springer Nature Singapore Pte Ltd. 2020   151
S. K. Ghosh et al. (eds.), *Recent Trends in Waste Water Treatment and Water Resource Management*, https://doi.org/10.1007/978-981-15-0706-9_15

past. To these locations, these withering waterbodies have been the most sustainable sources of water and have been the lifeline in arid/semi-arid regions of India. Traditionally in India, these man-made water bodies were fed with only rainwater (run-off) and were interconnected as a cascade. Water collected was used for both irrigation and domestic purposes throughout the year. However, due to the changing land-use pattern, diversion of valley systems and climate triggered weather patterns, loss of interconnectivity, catchment area changes and point discharges of sewage and human wastes, these water bodies have lost their capability for self-purification and water is far from reusable quality.

As indicated above, any new sewage generated from providing total sanitation will need to be treated and reused to meet sustainability goals. National programs in India emphasize this. Thus, while restoring water bodies at locations with a significant change in land-use patterns, including urbanization, we need to treat such decentralized wastewater in a passive and low energy approach. We seek to evolve new ways to treat these two types of wastewater. Firstly, evolve a treatment system that can work starting from a household (HH) level in rural, peri-urban and urban locations (daily discharge 250–400 l) to 1MLD levels for preventing water bodies from eutrophication, function as pre-treatment methods before the wastewater is returned to lakes and man-made water bodies. Firstly, the possibility of using this passive two-stage process is discussed for maintaining a water body for which run-off alone has become insufficient to hold enough water on a yearly basis. To solve this problem, we attempt to top-up the evaporative losses by using treated/partially treated sewage as water source with a passive treatment system constructed within the lake body. A new passive treatment technology is required which can overcome difficulties encountered while installing traditional STPs. Activated sludge process and SBR are most commonly applied treatment technologies in India but their high economic demand (~ Rs. 34–73/kL CPCB 2007) depending on level of treatment achieved, sludge production and disposal problems, technical expertise and no direct economic return make them inviable to adopt. Thus, conventional processes are not ideal for this purpose. Yet, natural water bodies remove pollutants through a combination of physical, chemical and biological processes (Skurlatov 1988; Koronelli 1996). Our main objective is to focus on biological energy and material transfer between different trophic levels and their role in treatments. The general trend is anthropogenic wastewater or agricultural run-off extended anoxic/anaerobic zone-phytoflagellates/ciliates/protists—algae—rotifers/cladocerans/snails/crustaceans—fishes—birds. In earlier publication, we proposed a passive treatment plant which would work on the principle of natural purification of water bodies while treating incoming sewage in 2–3 days to irrigation water quality by facilitating and optimizing each stage of natural purification.

In this proposed novel process, functional roles will be discussed for the first two stages using from experimental data (1L). First unit or stage is an anaerobic system, being most efficient carbon removal systems with approximately 70–80% of removed soluble/particulate carbon being converted to the gas phase. The by-product gas (~0.11–0.3 m^3/kg BOD) is an energy source when if collected can make the treatment economically sustainable. Sludge production is also less (2–10% of BOD removed) due to the lower growth and energy transfer in anabolic activities of anaerobic bacteria. Following this in the second stage, phytoflagellates, ciliates and protists which engulf suspended bacterial mass to finally reduce residual BOD and turbidity to levels <20NTU. Phytoflagellates are most abundant eukaryotic plankton in freshwater. Mixotrophic phytoflagellates (e.g. Euglenoids, Dinoflagellates) are known to provide both functions of protozoa and algae and are crucial actors in maintaining balance in biological system. Apart from being photosynthetic and using inorganic carbon to give off oxygen, they eliminate excess bacteria (Bird and Kalf 1987), cyanobacteria (Cole and Wynne 1974) and other algae, promote flocculation (e.g. Ochromonas, Aaronson 1973a, b) and help to decrease turbidity of effluent as well as BOD and suspended matter (Gerardi 1995; Curds 1968). These organisms have also been reported to release exoenzymes for extracellular digestion especially for flocculated and colloidal particles (Sanders and Porter 1988). Clearance rates (through both by ingestion and aid of enzymes) have been noted to range from 10^{3-4} bacteria/ml h. During some of laboratory experiments, approximately 7–77 bacteria have been found ingested per flagellate per hour depending on species (Sanders and porter 1988).

These studies suggest that mixotrophy (utilization of particulate food and dissolved organic matter), as well as photosynthetic fixation of organic matter in phytoflagellates, is a balanced process that can be regulated by environmental conditions possibly light or availability of soluble nutrients (nitrogen and phosphorus or vitamins) or both. Many observational and laboratory studies have reported of phagotrophy in phytoflagellates but so far no attempt has been made to use multiple consumption mode of these unique organisms in wastewater treatment. In this study, we have employed only two stages namely anaerobic digestion followed by the protozoan reactor to demonstrate a novel technique for in-line treatment of sewage to restore water bodies or to be applied as passive decentralized small-scale sewage treatment plants.

2 Materials and Methodology

Synthetic wastewater was used to study the treatment capability of anaerobic unit (4L) and protozoa (1L). The composition of media was kept constant throughout the experiment apart from during shock load (near 70d). Composition of media was 0.4 g/L—rice powder, 0.25 g/L—milk powder, 0.05 g/L—urea and 0.025 g/L—KH_2PO_4 with estimated COD of 640 mg/L, emulating typical sewage. The operating mode was fed-batch (fed twice in a day).

Filterable and total COD measurements were made for effluent from anaerobic reactor. MLVSS was estimated periodically. In order to understand the functional role of protozoa (1L), reactors were set up and operated in a similar mode as that of anaerobic system. The input for these reactors was effluent from anaerobic reactor described above. As light is an important factor, illumination with white LED was arranged (2 k lux) and nutrient reductions were monitored in both light and dark conditions. All the testing methods were from APHA. For stage 1, SVI, MLVSS through gravimetry, turbidity and COD are reported, whereas for second unit TKN with titrimetry, total phosphorus through persulfate digestion, nitrate, nitrite through cadmium reduction and turbidity along with COD were measured for stage two.

3   Results and Discussion

Stage 1: Anaerobic digestion has been applied for the treatment of various industrial and household wastewaters. The reactor set-up was operated in up-flow mode. The wastewater was introduced at the bottom of the tank, where sludge bed was maintained. The operation was at 27% (v/v) sludge volume with SVI of 27 ml/g VSS. The MLSS at time of operation was 6.4–8.2 g/L. Figure 1 shows total COD reduction for 75 days. While conducting this experiment, the reactor was already in steady-state condition. To establish functionality and kinetics of total reduction, total load was always kept below 1 kg COD/m^3 d. (~0.6 kg COD/m^3 d), which emulates sewage wastewater. To understand disturbances such as overload, from 70th day, load is increased to ~1.4 kg COD/m^3 d. This sudden load caused sludge to come out and a sudden dip in total COD reduction. Although within 5d, reactor had started to stabilize again. This will be monitored further for behaviour under different types of shocks.

Average COD reduction achieved so far has been 55% with 0.79/d of decay rate constant. Nitrogen and phosphorus removal in the reactor is >10%; thus, data is not shown here. The average turbidity of effluent from this stage was 110–180 NTU.

Stage 2: Traditionally, application of anaerobic technology is only as a pretreatment due to negligible removal of nutrients such as nitrogen and phosphorus and the presence of these cause high algal growth when disposed of directly and restrict its reusability to only agricultural lands. Phytoflagellates and protozoa due to

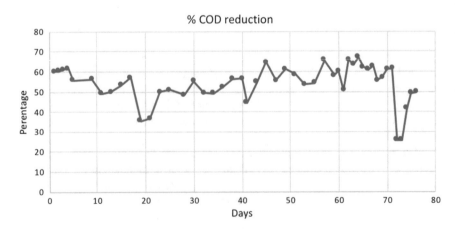

Fig. 1   COD reduction of 75 days in stage 1 anaerobic sludge bed reactor

capability of exhibiting multiple growth modes (Unpublished data, CST 2018) can remove a significant level of these nutrients.

The effluent from the first reactor was used as input of protozoan/phytoflagellate reactor. Figures 2, 3 and 4 show per cent and average removal of nutrients, COD and turbidity. To understand different modes and partition, the reduction levels between heterotrophic/phagotrophic and autotrophic/phagotrophic, experiments were conducted in light and dark. To the best of our knowledge, this has been not attempted before. The light requirement of phytoflagellates is significantly low, and it was necessary to maintain light limitation so as to maintain stable condition for them. Thus, 2000 lx (~30 $\mu M/m^2$ sec) of white LED illumination was provided. Although other intensity of lights was studied, this seemed to give the best results (Unpublished data, CST 2018).

In dark, due to the lack of photosynthesis, only two modes of nutrition can be functional. Either residual carbon source in particulate or soluble form is taken up or

	Parameter	% reduction
Light	COD	30
	TN	50
	TP	31
	Turbidity	70
Dark	COD	20
	TN	13
	TP	17
	Turbidity (%)	52

Fig. 2 Removal capabilities of phytoflagellates in light and dark in stage 2 reactor

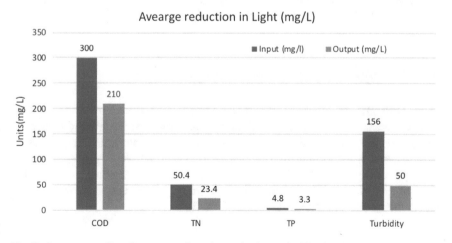

Fig. 3 Average overall performance and nutrient reduction under illumination

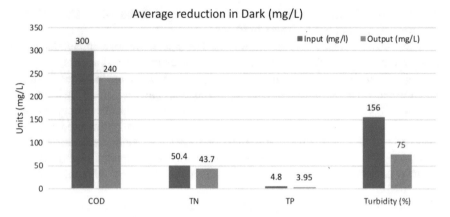

Fig. 4 Average overall reduction in dark

bacteria are being engulfed. In both cases, turbidity and COD from water are removed as shown in Fig. 3. However, the most interesting part is removal of additional COD and turbidity under light, which is counterintuitive as during illumination, inorganic carbon can also be fixed and total COD is expected to increase. This needs further research.

Apart from being an effective disinfectant, there is significant nitrogen and phosphorus removal as well (50 and 30%, respectively). This will ultimately reduce nutrient and C load on the discharge side. Adverse environmental factors such as low light, high bacterial turbidity and particulate forms of nutrients (nitrogen/phosphorus or vitamins) are some of the factors which might act as a switch between protozoa/phytoflagellates and chlorophyceae algae. Phagotrophy and mixotrophy as a source of nutrients have been reported for several algae species of Chrysophytes, Dinophyta and Cryophytes (Provasoli 1958; Aaronson 1980). This ability can be used in post-treatment of anaerobic effluent (Fig. 4).

4 Combined Nutrient Balance in Both the Units

When both units are operated sequentially, the expected reduction 70% COD, 55% total nitrogen, 25% phosphorus and most importantly ~80% turbidity reduction. Using this, technology enables it to be energetically independent with no aeration requirements. Even illumination requirements, which are very high in any algae tertiary treatment, are reduced and post-water is less turbid, bacteria-free and soluble nutrient-rich which can be directly used in flushing, gardening or agricultural fields and in the case of water body restoration (Fig. 5).

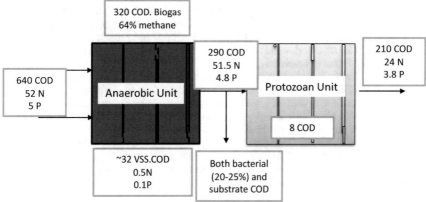

All units are expressed in mg/L.

Fig. 5 Flow of nutrients from both reactors and their functionality combined. Although some COD is present in the overall effluent, water is clear (<20NTU) and hardly any biologically useful C

5   Conclusion and Remarks

Difficulty in the operation of anaerobic reactors is to maintain stable conditions and mediating effects of overloads, the presence of inhibitors, changing abiotic factors (temperature). Early detecting indicators of any disturbance are necessary, so that reactors can be operated normally before its efficiency is reduced or methanogenic activity and biomass are washed off. A 27% v/v sludge in the first reactor is expected to overcome these issues. For the second unit, much less is known about the pathways in which organisms are able to function in light. Long-term studies on sludge stability and different photoperiods also need to conduct for this process to be established as stable. Apart from that, what are the other species which have the capability of performing dual roles and how these systems can be optimized is the future scope of this novel treatment process.

References

Aaronson, S. (1973a). Particle aggregation and phagotrophy by ochromonas. *Archiv fmikrobiologie, 92*(1), 39–44. https://doi.org/10.1007/bf00409509.

Aaronson, S. (1973b). Effect of incubation temperature on the macromolecular and lipid content of the phytoflagellate ochromonas danica. *Journal of phycology, 9*(1), 111–113. https://doi.org/10.1111/j.0022-3646.1973.00111.x.

Aaronson, S. (1980). Descriptive biochemistry and physiology of the chrysophyceae (with some comparisons to prymnesiophyceae). Academic Press New York.

Bird, D. F., & Kalff, J. (1987). Algal phagotrophy: regulating factors and importance relative to photosynthesis in dinobryon (chrysophyceae)1. *Limnology and Oceanography, 32*(2), 277–284. https://doi.org/10.4319/lo.1987.32.2.0277.

CPCB, 2007. (2006). Retrieved from http://cpcb.nic.in/cpcbold/upload/annualreport-06–07.pdf.

Cole, G. T., & Wynne, M. J. (1974). Endocytosis of microcystis aeruginosa by ochromonas danica. *Journal of Phycology, 10*(4), 397–410. https://doi.org/10.1111/j.1529-8817.1974.tb02732.x.

Curds, R. C. (1968). An experimental study of the role the ciliated protozoa in the activated-sludge process. *Water Pollution Control, 67,* 312–329.

Gerardi, M. H. (1995). Troubleshooting the sequencing batch reactor.

Koronelli, T. V. (1996). Principles and methods for raising efficiency of biological degradation of hydrocarbons in the environment. *Applied Biochemistry and Microbiology, 32,* 579–585.

Provasoli, L. (1958). Nutrition and ecology of protozoa and algae. *Annual Review of Microbiology, 12*(1), 279–308. https://doi.org/10.1146/annurev.mi.12.100158.001431.

Sanders, R. W., & Porter, K. G. (1988). *Phagotrophic phytoflagellates* (pp. 167–192). Springer, Boston, MA. https://doi.org/10.1007/978-1-4684-5409-3_5.

Skurlatov, Y. I. (1988). Fundamentals of the management of quality of water. *Ecological Chemistry of Aquatic Environment, 1,* 230–255.

Design and Characterization of Cold Plasma Ozonator for Wastewater Treatment

Harsha Rao, Lakshminarayana Rao, Haritha Haridas, D. K. Manju, S. Swetha and Hoysall Chanakya

Abstract As water scarcity becomes more severe, the need for more sustainable and holistic approaches to deal with the limited freshwater becomes more important. Developing decentralized technological solutions for grey water recovery from households and small communities is a potential way forward to reduce freshwater consumption. Towards this, in this study we report the design and characterization of a small-scale ozonator. The ozonator developed is used to degrade laundry grey water coming from consumer homes, SDBS and methyl orange. Ozonation of grey water from the washing machine resulted in 68.7% chemical oxygen demand (COD) reduction in 1 h of treatment at 1.2 g/h of ozonation. Ozonation of anionic surfactants (SDBS) and methyl orange resulted in 66.4% and 91.9% reduction in their concentration, respectively, over a period of 120 min of treatment. The kinetics studies show that the ozonation of grey water and SDBS follows first-order kinetics.

Keywords Grey water · DBD ozone generator · SDBS · Methyl orange · International society of waste management · Air and water

1 Introduction

Water scarcity is a universal problem. The UN Sustainable Development Goal 6 "Ensure access to water and sanitation for all" recognizes the right of citizens to access safe water and sanitation. Water scarcity, poor water quality and inadequate sanitation have a measurable negative impact on energy, public health and food security. By 2050, it is predicted that one in four people are likely to live in a country affected by chronic or reoccurring shortages of water. Indian domestic water consumption, especial in urban localities, is increasing and is expected to triple from the current demand by 2050. Domestic wastewater treatment and reuse are an attractive option to water conservation and management.

H. Rao · L. Rao (✉) · H. Haridas · D. K. Manju · S. Swetha · H. Chanakya
Centre for Sustainable Technologies, Indian Institute of Science, Bengaluru, India
e-mail: narayana@iisc.ac.in

H. Chanakya
e-mail: chanakya@iisc.ac.in

© Springer Nature Singapore Pte Ltd. 2020
S. K. Ghosh et al. (eds.), *Recent Trends in Waste Water Treatment and Water Resource Management*, https://doi.org/10.1007/978-981-15-0706-9_16

Grey water is any type of domestic wastewater that comes from kitchen sinks, washing machines, baths and hand basins free from urine and faecal matter (Jefferson et al. 2000). Grey water can be easily treated right at its source of origin and can be recycled for secondary usages such as gardening, floor and car washing purposes. There have been several technologies for the treatment of grey water including simpler natural treatment systems such as vertical wetlands to more advanced and complex treatment options such as membrane bioreactors (MBR) (Merz et al. 2007). Choice of grey water treatment technologies depends on its robustness to handle variations in organic, inorganic and pathogen concentration in grey water influent and consistently produce an effluent of an appropriate and safe quality to meet the required water quality standards for reuse. Advanced oxidation techniques such as ozonation owing its inherent strength to effectively treat a wide range of grey water (Den and Zhao 2015).

Ozone is a three-oxygen containing molecule having an atomicity of three. Ozone is a powerful disinfecting and oxidizing agent. The oxidation potential of ozone is 2.07 V which is 152% more than the next powerful oxidizing agent chlorine which has an oxidation potential of 1.36 V. Ozonation is a water treatment technique which is based on diffusing ozone into a water column. Ozonation is an advanced oxidation process, which involves in situ production of very reactive hydroxyl radical species, which are known to attack a wide range of organic compounds and microorganisms. When ozone is bubbled into water, it reacts with inorganic and organic compounds either directly or indirectly by the formation of hydroxyl radicals and oxidizes electron-rich molecules (Ried et al. 2009). Ozonation finds application for both disinfection and degradation of pollutants, inorganic and organic included (Ried et al. 2009). Ozonation is widely used in drinking water treatment systems because of its ability to disinfect without adding any other chemicals and/or leaving behind any residual chemicals like chlorine. Ozone can be produced by pumping in enough energy to break the firm O_2 bond to form O. Ozone production by electric discharge (DBD) is the most effective, economical and dependable means of controlled ozone production. It involves discharging high voltage, milliamp currents, across a gap through which filtered dry air or pure oxygen is flowing. The reaction is initiated when free, energetic electrons in the plasma zone ionize the oxygen molecules to form ozone.

Historically, ozonation has been used as the sole disinfection step in drinking water systems. Drinking water ozonation systems are large-scale systems which are capable of producing 100's of kg of ozone per hour. Using ozonation in decentralized household grey water treatment systems requires small-scale, reliable, low-cost compact ozone generators which are easy to operate and have minimum maintenance.

In this work, we report on the design and operation of a small-scale ozone generator which can be used for treatment of grey water coming from a typical 4-member household. The ozone throughput has been characterized for ozone production capacity, and its application in treatment of grey water from washing machines, degradation of SDBS and methyl orange has been studied.

1.1 Design of Dielectric Barrier Discharge (DBD) Plasma Ozone Generator

Dielectric barrier discharge (DBD) is a promising method of producing non-thermal plasma, which is widely used in a variety of industrial applications including ozone generation. Advantages of the DBD are its simplicity of its operation, capability to work at near room temperature and pressure and ability to work with very simple off-the-shelf commercial power supplies. In this work, a cylindrical electrode design ozone generator has been designed as shown in Fig. 1. In this design, an aluminium hollow cylinder, having 6 cm ID, 12 cm length and 3 mm wall thickness acted as one of the electrodes. An aluminium wire of diameter 3 mm and length 2.5 m wound as a coil formed the second electrode. A glass tube of diameter 6.5 cm ID, 14 cm length and 2 mm wall thickness placed in between these two electrodes formed the dielectric. A leak-tight acrylic casing, 10 cm OD and 15 cm length, with one air inlet port and one ozone outlet port, held the electrode assembly together. Compressed air was passed through the ozone generator. Many reactive species are produced during the plasma discharge in the air. In the gas phase, depending on the operating power level, nitrogen and oxygen molecules are dissociated by the energetic electrons in the discharge mostly leading to the formation of radicals (O and OH), ozone (O_3), nitrogen oxides (NO, NO_2), and excited N_2 through several reaction pathways (Sivachandiran and Khacef 2016).

Fig. 1 Co-axial DBD plasma ozone generator

When compressed air is passed through the plasma zone, various reactions are initiated. When free, energetic electrons interact with the compressed air, they dissociate oxygen molecules as shown in Eq. 1:

$$e^{-1} + O_2 \rightarrow 2O + e^{-1} \tag{1}$$

Following this, ozone is formed by a three-body collision reaction:

$$O + O_2M \rightarrow O_3 + M \tag{2}$$

M is a collision partner. It can be O_2, O_3, O or N_2.

1.2 Wastewater Treatment Using Ozonation

Ozone is known to completely oxidize organics to CO_2 and water either through direct reaction or through the formation of hydroxyl radicals. Oxidation reactions initiated by ozone in water are generally complex in nature, only part of the ozone reacts directly with dissolved solutes. Another part may decompose before reaction. Such decompositions are catalyzed by hydroxide ions and other solutes. Highly reactive secondary oxidants, such as OH˙ radicals, are thereby formed (Wang 2012).

The reactions of ozone with water include

$$O_3^- \rightarrow O^- + O_2 \tag{3}$$

$$O^- + H_2O \rightarrow OH˙ + OH^- \tag{4}$$

Ozone oxidizes electron-rich molecules attached to carbon–carbon double bonds such as aromatics and splits the long-chain compounds making them biodegradable.

In this study, anionic surfactants such as sodium dodecylbenzene sulphonate (SDBS) are chosen for the ozone degradation studies because of its widespread use in numerous detergents and other household cleaning products. It belongs to the group of surfactants called linear alkyl benzene sulphonates (LAS). The average SDBS concentration in domestic wastewater ranges between 1 and 10 mg/L (Zhang et al. 1993). Conventional oxidants such as hydrogen peroxide and potassium permanganate have little effect on the removal of LAS, and when chlorine is used, low yields of chloroform; a known carcinogenic compound is formed. Ozonation of SDBS in aqueous solution consists of a free radical mechanism, in addition to direct reactions of molecular ozone. A general mechanism of the ozonation of SDBS can be summed up as shown in Equations 5 and 6.

Direct reaction:

$$NaDBS + zO_3 \rightarrow intermediates \rightarrow final\,products\left(CO_2,\ SO_4^{2-}, H_2O\right) \tag{5}$$

Free radical reaction:

$$NaDBS + OH^{\cdot} \rightarrow intermediates \rightarrow final\,products\left(CO_2,\ SO_4^{2-}, H_2O\right) \qquad (6)$$

The decolourization of methyl orange by ozonation was also studied in this work. Azo dyes are widely used in textile and paper industries. Disposal of dye-rich wastewater from these industries is an environmental concern since the associated colour is quite noticeable and may have carcinogenic effects. When ozonation of an azo dye is conducted, the possible pathways are homolytic fission of the excited dye into radicals, electron transfer of the excited dye to form a radical dye cation, decomposition by superoxide radical anion and decomposition by singlet oxygen, resulting in a reduction in colour (Wang 2012).

2 Materials and Methods

2.1 Materials

A commercially available high voltage AC power supply system (Power: 80 W, Output: 10 kV 30 mA) was used as a power supply for DBD plasma generator. All chemicals (SDBS and methyl orange) were of reagent grade, and they were used as received. Stock solutions were prepared in distilled water.

2.2 Experimental Methods

The experimental set-up consists of a long cylindrical PVC pipe which acted as an ozone contact reactor, a co-axial DBD plasma ozone generator which produced ozone from compressed air, grey water circulating pump and spray nozzle system which increased the contact time between ozone and grey water and an air compressor, as illustrated in Fig. 2. Three types of studies, namely (1) laundry grey water treatment using ozonation, (2) SDBS destruction using ozonation and (3) methyl orange degradation studies, were all conducted in batch mode. The ozone throughput from the designed plasma ozonator was characterized using the wet KI method. The ozone throughput at different airflow rates was measured.

In a typical laundry grey water treatment batch experiments, 4 L of grey water to be treated is filled in the pipe. A mixed water sample obtained by mixing the discharge water from a household laundry washing machine's rinse cycle and wash cycle water in 1:1 ratio was used in this study. Compressed air at a known flow rate from the compressor is fed to the plasma generator, and the outlet from plasma generator is bubbled through the grey water. Ceramic aerators capable of producing fine bubbles were used as submerged aerators. System was provided with a recirculation

Fig. 2 Experimental set-up for ozonation of grey water. (i) DBD plasma generator, (ii) ozone inlet, (iii) batch reactor, (iv) air compressor, (v) water circulating pump

pump fitted with a nozzle spray to maximize contact time between ozone and grey water. Water samples were collected at regular frequency over the entire duration of ozonation which was up to 60 min. Parameters such as COD, TOC and pH of the grey water were measured using standard procedures. Attempts were also made to determine the kinetics of the COD reduction due to ozonation.

In a typical SDBS destruction batch experiments, 1L of water having an initial concentration of 7 ppm of SDBS was taken in a batch reactor and the ozone was bubbled through. Water samples at regular intervals were taken and analysed for SDBS concentration and pH as a function of ozonation time. The concentration of SDBS was determined using methylene blue active substances (MBAS) method. Attempts were also made to determine the kinetics of the SDBS reduction due to ozonation.

To study the methyl orange degradation, 1 L of water having an initial concentration of 500 ppm of methyl orange was taken in a batch reactor and the ozone was bubbled through the water. Water samples at regular intervals were taken and analysed for methyl orange concentration and pH as a function of ozonation time. The concentration of methyl orange was determined using UV spectrophotometer.

3 Results and Discussion

3.1 Ozone Throughput Studies

Figure 3 shows the ozone throughput as a function of inlet airflow. As can be seen from Fig. 3, increasing airflow rate increased ozone throughput. For 20 lpm of airflow rate, 1210 ± 10 mg/h of ozone throughput was obtained, and at 100 lpm of airflow

Fig. 3 Ozone throughput (mg/h) as a function of airflow rate (lpm) for co-axial DBD plasma reactor

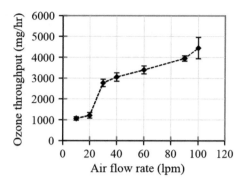

rate, the ozone throughput measured was 4500 ± 500 mg/h of ozone. We believe that increase in airflow rate leads to higher plasma volume in the reactor leading to higher ozone throughput.

3.2 Treatment of Grey Water Using Ozonation

The feed grey water, which was a mixed sample collected from a household laundry machine, had an initial COD in the range between 500 and 900 mg/L. A 4 L sample was treated using ozonation at 20 LPM of airflow rate. Under these conditions, we noticed that ozonation resulted in 68.7% reduction in COD reduction in the first one hour of treatment at 1.2 g/h of ozone flow rate. During experiments, we noticed that though the grey water coming from the laundry machines created foam during initial 10–15 min, it quickly stopped showing the chemical degradation of surfactants present in the grey water. Figure 4 shows the visual appearance of grey water as a function of ozonation time. As shown in Fig. 4, with ozonation, the turbidity of the water reduced and at the end of 60 min of ozonation, the water looked clear. As illustrated by Fig. 5, in 60 min of treatment the COD reached the effluent discharge limit of 250 ppm in all the treated samples regardless of initial value of COD. The pH remained alkaline, between 8 and 9, during the treatment of laundry grey water.

Fig. 4 Reduction in turbidity of laundry grey water over the period of ozonation

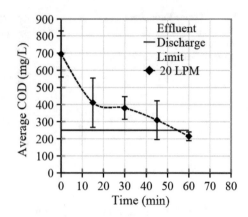

Fig. 5 Reduction in chemical oxygen demand (COD) as a function of time (min) for 4 L of ozonated washing machine discharge at 20 lpm airflow rate

The laundry grey water which typically is rich in SDBS surfactants gets reduced through the hydroxyl mechanism of oxidation as discussed in Sect. 1.2. Due to its electrophilic nature (electron preferring), the OH^- can non-selectively oxidize almost all electron rich organic molecules including SDBS.

Total organic Carbon (TOC) was also measured using a Shimadzu TOC-5000A unit by combustion catalytic oxidation method and detection using an infrared gas analyzer. The TOC for ozonated laundry water reduced from 69.1 to 28.2 ppm in 60 min of treatment.

The first step in modelling the process for the study of the kinetics of wastewater ozonation is the determination of the rate constant of the ozone reactions. The kinetic studies of wastewater ozonation are shown in Fig. 6. As can be seen in Fig. 6, we can see that the reduction of COD follows first-order kinetics with respect to ozone and solute concentration at 8.8 pH. Table 1 shows the rate constants of the destruction of various compounds with ozone.

Fig. 6 In (Co/C) versus time (min) for 4 L of ozonated washing machine discharge at 20 lpm airflow rate

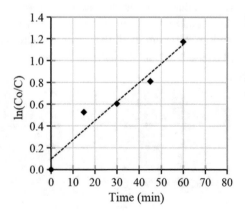

Table 1 Reaction rate constants of ozone

Component	pH	k_{O3} (s^{-1})
COD	8.8	0.0197
SDBS	8.5	0.0213

3.3 Removal of SDBS Using Ozonation

SDBS concentration gradually reduced to 66.4% of the initial amount over a period of 120 min as shown in Fig. 7. We believe that this observed reduction is mainly due to its reaction with OH⁻ radical. The solution remains alkaline over the period of ozonation. Alkaline conditions favour the ozonation rate of the SDBS compound, since hydroxyl radical reactions are the main route of its degradation. It follows first-order rate kinetics as shown in Fig. 8 and Table 1.

Fig. 7 SDBS degradation by ozonation as a function of time (min) at 20 lpm airflow rate

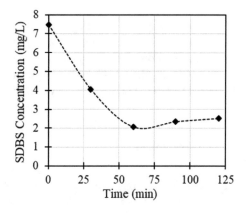

Fig. 8 In (Co/C) versus time (min) for SDBS ozonation at 20 lpm of airflow rate

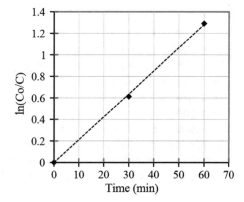

Fig. 9 Methyl orange
degradation by ozonation as
a function of time (min) at
40 lpm airflow rate

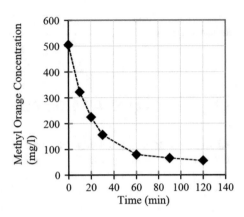

3.4 Methyl Orange Degradation by Ozonation

The decolouring of 1 L methyl orange solution was studied using DBD plasma generator at 40 lpm (3.05 g/h of Ozone) of airflow rate as illustrated in Fig. 9. The dye concentration of 500 ppm with the initial pH of 6.7 was fixed. Experimental results confirm that 91.9% reduction in methyl orange colour concentration was observed over a period of 120 min of ozonation.

4 Conclusions

This study has demonstrated that a small-scale reliable ozone generation system can be developed and can be used to treat domestic grey water. The following can be concluded:

- Ozonation of grey water from washing machine resulted in 68.7% and 59.2% reduction of COD and TOC, respectively, in 1 h of ozonation, regardless of the initial COD value of the sample.
- Turbidity of samples decreased to a significant extent over the course of ozonation. Foaming also reduced after treatment which indicates ozonation causes surfactant degradation in grey water.
- Ozonation of anionic surfactants and methyl orange resulted in 66.4% and 91.9% reduction in concentration, respectively, over a period of 120 min of ozonation. The solution remained alkaline over the period of ozonation.
- Ozonation is highly recommended due to its degradation effectiveness, rate of degradation, relatively simple experimental design and scalability.

References

Den, Y., Zhao, R. (2015). Advanced oxidation processes (AOPs) in wastewater treatment. *1*(3), 167–176.

Jefferson, B., Laine, A., Parsons, S., Stephenson, T., & Judd, S. (2000). Technologies for domestic wastewater recycling. *Urban Water, 1,* 285–292.

Merz, C., Scheumann, R., El Hamouri, B., & Kraume, M. (2007). Membrane bioreactor technology for the treatment of greywater from a sports and leisure club. *Desalination, 215,* 37–43.

Ried, A., Mielcke, J., & Wieland, A. (2009). The potential use of ozone in municipal wastewater. *Ozone Science and Engineering, 31*(6), 415–421.

Sivachandiran, L., & Khacef, A. (2016). In situ and ex situ NO oxidation assisted by sub-microsecond pulsed multi-pin-to- plane corona discharge: The effect of pin density. *RSC Advances, 6*(36), 29983–29995.

Wang, J. L., & XU, L. J. (2012). Advanced oxidation processes for wastewater treatment: Formation of hydroxyl radical and application. *Critical Reviews in Environmental Science and Technology, 42*(3), 251–325.

Zhang, C., Valsaraj, K. T., Constant, W. D., & Roy, D. (1993). Aerobic biodegradation kinetics of four anionic and nonionic surfactants at sub- and supra-critical micelle concentrations (CMCs). *Water Research, 33,* 115–124.

Natural Adsorbents for Removal of Different Iodine Species from Aqueous Environment: A Review

Jakkapon Phanthuwongpakdee, Sandhya Babel and Tatsuo Kaneko

Abstract Radioactive iodine in the nuclear waste can cause adverse effects on the human and other living organisms. I-129 and I-131 are released into the aqueous environment via the nuclear accident, weapon testing, scientific procedures as well as hospitals. The radioactivity of these substances is fatal to human and other living creatures. This study reviews different types of natural adsorbents that can be utilized for the application of radioactive iodine remediation in the water. To solve the problem, scientifically, the conventional anion-exchange resin is typically employed for the anionic species of iodine, but it is deemed unsustainable due to dependency over petroleum-based chemical. As a result, there is a need for the search of more sustainable adsorbents, which include activated carbon, soil and clay materials, mineral and even polysaccharide. The activated carbon, when impregnated with silver, can adsorb up to 126.9 mg/g of I^-. In its fiber form, the adsorbent has the equilibrium capacity of 305 mg/g. Clay can be modified to contain quaternary amine functional group and has I^- coefficient sorption, K_D, value of 297–9610 mL/g. For the mineral, many kinds had been tested for iodine removal. Non-modified hydrotalcites can reach I^- adsorption equilibrium of 192.5 mg/g in 1 day, while the magnetite–imidazole-coated silica adsorbed 130 mg/g within 40 min. The cyclodextrin–chitosan complex has one of the highest I_2 adsorption capacities of 1200 mg/g. The information shows that natural adsorbents can be considered as a green alternative to the conventional anion-exchange resin, and more studies to optimize these materials are needed in the future.

Keywords Radioactive iodine · Activated carbon · Soils · Clay · Minerals · Polysaccharides · Adsorption efficiency · K_D

J. Phanthuwongpakdee · S. Babel (✉)
School of Bio-Chemical Engineering and Technology, Sirindhorn International Institute of Technology (SIIT), Thammasat University, P.O. Box 22, 12121 Pathum Thani, Thailand
e-mail: sandhya@siit.tu.ac.th

J. Phanthuwongpakdee · T. Kaneko
Graduate School of Advanced Science and Technology, Japan Advanced Institute of Science and Technology (JAIST), 1-1 Asahidai, 923-1292 Nomi, Ishikawa, Japan
e-mail: kaneko@jaist.ac.jp

© Springer Nature Singapore Pte Ltd. 2020
S. K. Ghosh et al. (eds.), *Recent Trends in Waste Water Treatment and Water Resource Management*, https://doi.org/10.1007/978-981-15-0706-9_17

1 Introduction

Radioactive iodine contained in the nuclear waste can create adverse effects on human and other living organisms. Of all the iodine isotopes, I-129 and I-131 should be the most concerned in term of environmental impact and human health. With its 8 days of decay half-life, Iodine-131 (I-131) is described as the most hazardous radionuclide released into the environment (Morita et al. 2010). Similarly, I-129 can also create such adverse health effect to human. It has a lower activity per ton of fuel, but possess a potential to create a long-term risk and is a major problem related to waste disposal (Denham et al. 2009). I-129 exists in a sufficient quantity, and it has a much longer half-life time at 1.6×10^7 years. During the time of the large-scale accident, I-131 might be a more problematic species due to its higher radioactive energy while the accumulation of I-129 in the groundwater and soil has been observed around the nuclear waste disposal sites such as Savannah River and Hanford Site (Denham et al. 2009).

It is known that radioactive element is created by cosmic radiation and the natural disintegration series of other elements (Kohman and Edwards 1966). On the other hand, it is also produced by anthropogenic activities such as weapons testing, science experiment, medical procedures, and the fission process in nuclear power plants. Table 1 shows the major sources and incidents that have caused the released of radioactive iodine. Creating energy from the fission process in nuclear power plants seems to be one of the promising ways of securing electrical power in the future. However, since the first nuclear power plant was constructed, the safety concern relating to its waste products, which are being released into the environment, has been the issue to be dealt with (Ramana 2009). Chernobyl disaster occurred in 1986 with the explosion of reactors in Chernobyl Nuclear Power Plant located in Pripyat, Ukraine. Following the Earthquake near Japan in March 2011, the tsunami destroyed parts of Fukushima Nuclear Power Plants, releasing the substance from the reactors into the environment. Savannah River Site, Hanford Site, Nevada Testing Site, and nuclear reprocessing plants are adding amounts of radioactive iodine into the water resources as well.

When comparing the natural products to those of the quantity released by human actions, it can be seen that human contributed to a much higher amount of radioactive iodine (Kohman and Edwards 1966; Hu et al. 2010). According to Toxicological Profile for Iodine, radioactive iodine can destroy thyroid gland and increase the risk for hypothyroidism, thyroid nodules, and cancer (Agency for Toxic Substances and Disease Registry 2004). Human's thyroid gland, naturally, adsorb iodine for the purpose of hormonal, growth, and metabolic regulation. It can contain up to 30–40 times of iodine concentration than that of in the bloodstream (Braverman and Cooper 2005). Moreover, the pathway of iodine includes the adsorption from the gastrointestinal tract and lungs which the substance eventually ends up in the bloodstream (Agency for Toxic Substances and Disease Registry 2004). Iodine can also be found in other parts of human body including gastric mucosa, salivary glands, mammary glands, choroid plexus, ovaries, placenta, breast tissues, breast milk, and

Table 1 Amount of radioactive iodine released from different kinds of sources and incidents

Source/incident	Environment affected	Radioactive iodine	Concentration or discharge	Total activity	Reference
Chernobyl, 1986	Soil and atmosphere	I-129	–	0.04 TBq	United Nations Scientific Committee on the Effects of Atomic Radiation (2000)
Chernobyl, 1986	Soil and atmosphere	I-131	–	1.76×10^6 TBq	United Nations Scientific Committee on the Effects of Atomic Radiation (2000)
Fukushima, 2011	Air	I-131	–	153–160 PBq	Kawamura et al. (2011)
Fukushima, 2011	Ocean	I-131	300–4000 Bq/L	3–11 PBq	Kawamura et al. (2011)
Fukushima, 2011	Ocean	I-129	14×10^7 to 62×10^7 atoms/L	–	Hou et al. (2013)
Savannah River Site	Groundwater	I-129	2.1–59.9 Bq/L	–	Kaplan et al. (2014)
Nevada Testing Site	Groundwater	I-129	0.777 Bq/L	–	Smith et al. (2003)
Harford Testing Site	Groundwater	I-129	2.2–42.5 pCi/L	–	Zhang et al. (2013)
Reprocessing Plants	Baltic Sea and rivers nearby	I-129	0.71–13.80 atoms/L	–	Keogh et al. (2007), Aldahan et al. (2006)
Royal Hospital, Muscat, Oman (2006–2009)	liquid waste in the hospital	I-131	–	3777 MBq	Ravichandran et al. (2011)
Medical institutes, Japan	Tap water	I-131	13–15 Bq/kg	–	Kamei et al. (2012)

skin (Jhiang et al. 1994; National Academy of Science 1999). The exposure pathway of radioactive iodine to human includes the ingestion of cow milk, fruits, vegetable, fish, water, and inhalation (National Academy of Science 1999; Peterson et al. 2007).

A large amount of iodine in the ocean can also create negative feedback to the earth's ozone layer. The iodine in the sea and marine aerosol are important aspects of the earth's temperature and sun radiation regulation (O'Dowd et al. 2002). Above the ocean, 25% of ozone chemical had been lost due to the emission of iodine and 70% increase in ozone depletion due to iodine is estimated from that of in pre-industrial era (Prados-Roman et al. 2015). High accumulation of long-lived I-129 in the ocean can contribute to such adverse effects.

With all the impacts mentioned, the scientific field has been challenged with the search for a suitable technology to remove and immobilize the pollutant, especially in the water environment. The technologies used for radioactive iodine removal in the water are summarized in Table 2. The techniques include membrane separation, electrolysis, chemical precipitation, phytoremediation, and adsorption. The membrane separation and electrolysis are mostly limited to the drinking water application. While these techniques could remove more than 90% of the radioactive iodine in various concentrations, the pre-treatment by adsorption is still needed to prevent the process failure (Hamasaki et al. 2014; Chae and Kin 2014). The drawback of chemical precipitation is that the addition of chemical such as polyaluminum chloride (PAC) and silver-containing agent is not suitable for the open environment (Kosaka et al. 2012; Eden et al. 1952). As a result, treatment plant is usually required, and transportation of contaminated water is thus required. Species of brown algae were proven to accumulate iodine, but the process takes time and further studies are required to control the rate of accumulation (Küpper et al. 1998; Nitschke and Stengel 2014). For the adsorption, the synthesized polystyrene-based adsorbent is commonly used. However, its dependence on petroleum-based chemical is deemed unsustainable.

In this report, the adsorption of radioactive iodine, iodide, and iodate is being considered. It reviews different kinds of natural adsorbents with and without modification that can be employed for adsorption of different iodine species. It also presents the optimum conditions for each adsorbent. The comparison of the adsorption capacities between each group and with ion-exchange resin is also reported. Thus, this review helps in identifying the potential natural adsorbents for the removal of iodine from water environment.

2 Principle of Adsorption

Adsorption is the process of the association or/and adherence of fluid phase on the surface of the solid phase. This can happen in two ways— through physical and/or chemical interactions. The physical adsorption usually deals with London dispersion forces (as one of Van der Waals forces). London dispersion force is an intermolecular attraction between uneven electrons and protons, creating temporary dipole. The

Table 2 Typical techniques used for the removal of radioactive iodine in the water

Types of technology	General principle of removal	Maximum iodine removal (%)	Initial concentration	Reference
Membrane separation	1. Pressure-driven—use of pressure to force the water through the membrane where the particulate is separated from the water, depending on the membrane porous size 2. Ion-exchange—use of membrane with a cationic functional group to trap the anionic iodine species	>95	365 kBq/L	Hamasaki et al. (2014), Sancho et al. (2006), Van Der Bruggen et al. (2003)
Coagulation–flocculation–sedimentation	Precipitation of iodine species with polyaluminum chloride (PAC) and silver nitrate in acidified environment	>95	150 μg/L	Kosaka et al. (2012), Eden et al. (1952), Kitada et al. (2015)
Electrolysis	Separating the anionic species of radioactive iodine by applying direct electrical current (DC)	99	4000 μg/L	Hamasaki et al. (2014)
Phytoremediation	Bioaccumulation and/or surface adsorption by species of algae and water plants	>60	1000 kq/L	Morita et al. (2010), Fukuda et al. (2014), Hosseini (2010)
Polystyrene-based anion-exchange resin	Anion-exchange and adsorption between anionic species of iodine and Cl^- or OH^- on the resin beads	>70 (2979.6 mL/g K_D and 5.9 μg/g maximum adsorption)	7.88 μg/L	Parker et al. (2014)

interaction energy depends on the surface contact area of the molecules, polarizability (based on molecule size) and the number of pi-bonding present. The chemical adsorption deals with ionic and covalent bonding between the target molecules and the functional group of the adsorbent. The physical adsorption is usually weaker than the chemical adsorption.

The binding process of adsorption is related to Eq. (1).

$$\Delta G = \Delta H - T \Delta S \tag{1}$$

where, ΔG is the change of Gibbs free energy expressed in kJ/mol and T is temperature. ΔG is associated with the adsorption process as shown in Eq. (2).

$$\Delta G = -RT \ln K \tag{2}$$

where, K is the thermodynamic constant related to the binding of the sorbate to the adsorbent. R is the gas constant. ΔG is related to the spontaneity of the adsorption process where the higher in negative value of ΔG reflects more energetically favorable adsorption (Plazinski 2012). Substituting Eq. (1) into Eq. (2) would yield Eq. (3).

$$\ln K = -\frac{\Delta H}{RT} + \frac{\Delta S}{R} \tag{3}$$

For an experiment with changing in temperature is observed, ΔH and ΔS can be obtained from the slope and the intercept of the plot $\ln K$ against $1/T$.

If we are determining the equilibrium relationship between solid and aqueous solution, we can first equilibrate the known adsorptive solution and the sorbent at a fixed temperature and pressure. The equilibrium relationship, described by isotherms, will be obtained when the adsorptive solution and the adsorbent are allowed to make contact for a period of time. The amount of adsorption or amount of mass of solute adsorbed per unit mass of adsorbent, q (expressed in mol/g or mg/g) can be calculated with the mass balance equation (Eq. 4).

$$q = \frac{(C_i - C_e)V}{m} \tag{4}$$

where, C_i and C_e are the initial and final adsorptive concentration, respectively. V is the volume while m is the mass of the adsorbent. To determine the most suitable adsorption model, it is important to understand that the adsorption isotherms exhibit many different shapes for a different combination of sorbates and sorbents. By plotting q against C_e, if a linear graph is obtained, the relationship between q/C_e can be expressed as a ratio of solid solution distribution coefficient K_D (Eq. 5).

$$K_D = \frac{q}{C_e} \tag{5}$$

Linear adsorption usually occurs at low solute concentration and low amount of sorbate (Delle Site 2001). If the plot of q against C_e yields a nonlinear graph, it indicates that after a period of time, the sites of sorbent become saturated and the degree of adsorption does not change anymore. Two kinds of models can be used for this particular case. The first model derived from the equation of the change in final concentration over time which expressed as Eq. (6).

$$\frac{\partial C_e}{\partial t} = -k_1 m \Delta C_e (q_{max} - q) + k_2 q m \qquad (6)$$

where q_{max} is the adsorption capacity or adsorbed solutes concentration corresponding to the total number of surface sites. k_1 and k_2 are the rate constant for adsorption and desorption. The rate of adsorption is assumed to be proportional to C_e and to the difference in q_{max} and q. The equation can be reduced to express the relationship between q and C_e Eq. (7).

$$q = \frac{q_{max} K_L C_e}{1 + K_L C_e} \qquad (7)$$

where, K_L is the Langmuir constant or coefficient related to the enthalpy of adsorption. To get a linear relation between q and C_e, the Eq. (8) is used.

$$\frac{1}{q} = \frac{1}{q_{max}} + \frac{1}{K_L C_e q_{max}} \qquad (8)$$

where, q_{max} can be calculated from the inverse of the intercept. Therefore, K_L can be calculated with the inverse of slope multiply by q_{max} obtained.

In Langmuir model, the relation between q and q_{max} can be described as when the solute concentration is increasing at low surface coverage with the available adsorption sites reaching saturation and q is increasing linearly to approach q_{max}. The model has q_{max} representing monolayer coverage and assumes that adsorption energy is independent of degree of surface coverage (Delle Site 2001). It also assumes the energy to be equal for all sites, where only localized ones are involved in the adsorption process without interaction between adjoining sorbed molecules (Delle Site 2001).

The second model takes into account the degree of surface coverage and sorption energy for the sites. It is called Freundlich model where the relationship of q and C_e is expressed as Eq. (9).

$$q = K_F C_e^n \qquad (9)$$

where, K_F is used as the Freundlich coefficient and n is the exponent relates to adsorption intensity and free energy. To get the linear plot for Freundlich isotherms, log transformation is used (Eq. 10).

$$\mathrm{Log}q = n\mathrm{Log}C_e + \mathrm{Log}K_F \qquad (10)$$

Using the linear plot of Freundlich model, n and K_F can be obtained from the slope and intercept, respectively. If n is 0, a constant adsorption-free energy at all sorbate concentration is observed. On the other hand, a value of n more than 1 means the increase in the presence of sorbate enhances the free energies for further adsorption, but when n is less than 1, weaker free energies with increasing sorbate is observed.

It is worthy to note that the linear, Langmuir or Freundlich isotherms equation cannot solely explain the sigmoidal isotherm relation. Instead, sigmoidal behavior signifies the combination of two different adsorption mechanisms, one after another. It occurred due to alternation in the functional groups during the adsorption, numerous components of a heterogeneous sorbates or diversity in free energies related to the adsorption process.

3 Natural Adsorbents for Iodine

3.1 Activated Carbon

Iodine adsorption by activated carbon has been reported for several decades. The activated carbon is derived from the carbonaceous materials such as coconut, nutshell, wood, and lignin. Sometimes, it is made of coal and can be described as activated coal (for this report, the activated coal maybe refers as an activated carbon). The gaseous iodine adsorption capacity is typically used to characterize the commercial activated carbon in the form of "iodine number," which refers to how much amount of iodine (mg^2) is being adsorbed by a gram of activated carbon. The activated carbon is, therefore, regarded as a potential adsorbent for the radioactive iodine in the water. Activated carbon has been utilized and studied extensively for both air and water pollution adsorption because of its large surface area. Many studies have shown a successful iodine adsorption of radioactive venting from the nuclear power plant exhaust pipe and that of polluting the air environment (Adams and Browning 1960; Zhou et al. 2014). However, in this report, the only removal of iodine in the water is considered and summarized in detail.

The activated carbon adsorbs sorbates by London dispersion force. It consists of micro-, meso- and macropores structures, where the major adsorption of iodine occurs mostly at the micropores (Yang et al. 1993). As a result, it is critical how these pores are arranged depending on the form of the activated carbon (Barkauskas and Dervinyte 2004). The comparison of activated carbon in different forms is exhibited in Table 3. One study compared the ability of granular and fiber activated carbon to remove I_2 and I^- (prepared by HI), in the acetic acid (Yang et al. 1993). Under acidic environment, the activated carbon fiber could absorb 305 mg/g I_2 and 350 mg/g I^-, higher than the adsorption capacity of 195 mg/g I_2 and 205 mg/g I^- in the granular form of activated carbon (Yang et al. 1993). The outcome coincided with the BET

Table 3 Comparison of iodine adsorption with different types of activated carbon

Type of activated carbon	[a]Maximum iodine adsorption (mg/g)	Removal (%)	Equilibrium time	Experimental condition					Reference
				Iodine species	Initial iodine concentration (mg/L)	Adsorbent dosage (g/L)	Medium	pH	
Carbon fiber	305	–	2 h	I_2	800	–	Acetic acid	–	Yang et al. (1993)
Granular	195	–	24 h	I_2	800	–	Acetic acid	–	Yang et al. (1993)
Granular	0.635	87.50	–	I^-	1.015	1	DI water	8	Hoskins and Karafil (2002)
Granular	1.68×10^{-3}	18.50	–	I^-, IO_3^-, I_2	9.19×10^{-3}	1	Groundwater	8.2	Parker et al. (2014)
Granular	–	60	–	I^-	40×10^{-3} to 50×10^{-3}	–	Tap water	7	Sato et al. (2011)
1.05 wt% silver-impregnated carbon	~126.9	–	–	I^-	~1 to 200	1	DI water	5	Hoskins and Karafil (2002)
Silver-activated carbon aerogel	0.638	–	7 day	I^-	0.15	~0.2 (2000 g/L/min)	DI water	7	Sánchez-Polo et al. (2006)
Copper-modified activated carbon	1.91	95.7	2 h	I^-	2	1	DI water	1	Zhang et al. (2017)

[a]The value is obtained from the experimental data and not from the mathematical model

surface area for activated carbon fiber which is two times higher than that of the granular form. In fiber form, micropores were exposed to the surface of the sorbate, creating a fast iodine binding and diffusion rate (Yang et al. 1993). On the other hand, in granular form of activated carbon, there was a sequence binding and diffusion of iodine from the mesopores at the surface to micropores within the bulk of the sorbate (Yang et al. 1993).

Besides the physical form of activated carbon, its activation processes involved are important aspects in the adsorption mechanism. In the production phase, biomaterials would go through pyrolysis or oxidation for its physical activation to produce an activated carbon. In the pyrolysis process of activated coal where the temperature was increased to about 400 °C, iodine adsorption sites in coal were enhanced with increasing oxygen cross-link, and this results in higher adsorption capacity (Rodriguez and Marsh 1987). On the other hand, the oxidation process decreased the size of micropores and reduced the adsorption capacity in the activated carbon (Rodriguez and Marsh 1987).

In the laboratory experiment, the activated carbon could remove up to 87.50% of 2 mg/L iodide in the solution (Hoskins and Karafil 2002). However, studies have shown that the removal of iodine in other types of water medium may yield lower iodine removal. Compared to the results of the batch experiment using distilled water, the one with 50 μg/L I^- and IO_3^- in tap water showed 60% and almost 0% removal, respectively (Sato et al. 2011). On the other hand, the result of adsorption in groundwater was the lowest with about 20% maximum adsorption at 9.77 μg/L initial total iodine (Parker et al. 2014). The presence of other chemical species in groundwater environment may cause a competition on available sorption site, resulting in less removal percentage (Parker et al. 2014).

The activated carbon possesses several O_2-containing functional groups such as carbonyl, phenol, lactone, and carboxyl group (Barkauskas and Dervinyte 2004; Nwosu et al. 2009). It is still not certain which group, in particular, is responsible for the sorption of iodine. However, there was an attempt to find the relation between the number of these functional groups in the nutshell, quantified by the method of NaOH titratable surface and iodine sorption capacity (Nwosu et al. 2009). The sorption of iodine was found to be dependent moderately on the quantity of functional groups with a linear statistical correlation of 0.6115 (Nwosu et al. 2009). Isolating each functional group for the model would provide an answer to which O_2-containing functional group is responsible for this sorption process.

Many times, the porous carbon can be modified to contain the inorganic material such as silver and aluminum for different sorption specification. Silver-impregnated carbon is often used for iodine sorption, and it is proved to be able to adsorb more iodine. The sorption capacity value of 1.05% silver-impregnated activated carbon was nearly two times higher than that of virgin activated carbon (Hoskins and Karafil 2002). However, with only 0.05% of silver impregnation, there was no significant change in the value compared to that of virgin activated carbon (Hoskins and Karafil 2002). The silver-activated carbon aerogel was also brought to study in the dynamic adsorption test, where the results exhibited about 0.251–0.638 mg/g sorption capacity, depending on pH and porous characteristics of the aerogel (Sánchez-Polo et al.

2006). The nano-composite of Cu_2O/Cu-activated carbon was synthesized with the ability to remove 95.7% of 2 mg/L I^- and has about 1.91 mg/g adsorption capacity.

The sorption of iodine to activated carbon is affected by pH. At higher pH, activated carbon absorbed iodine less effectively as elevated amount of OH^- competes for the positive adsorption sites (Hoskins and Karafil 2002). The pH effect was demonstrated in the batch experiments of unmodified and copper-impregnated activated carbon. The results showed that all kinds of the activated carbon experience a lower I^- adsorption capacity as pH increases from 3 to 11, although the copper-impregnated activated carbon was shown to be more stable across the pH range (Sánchez-Polo et al. 2006).

The unmodified activated carbon may not be the best selection for the aqueous iodine adsorption because its efficiency decreases drastically when it is used in the natural environment. However, the modification can help to improve its performance. The carbon fiber and silver-impregnated activated carbon are among the best adsorbents in this category. In addition, the activation of biomaterials has brought about the potential in finding alternative and innovative way of making activated carbon. Several waste products such as coffee residue (Boonamnuayvitaya et al. 2004; Boonamnuayvitaya et al. 2005), soybean oil cake (Tay et al. 2009), and even cigarette butts (Sun et al. 2017) became possible raw materials for the activated carbon, and these ideas could be applied for iodine removal in the future.

3.2   Soil and Clay

Iodine is often found as a content of natural soil and clay. Many times the soil itself is contaminated by radioactive iodine (Mimides and Kathariou 2005; Miyake et al. 2012). However, based on the contamination occurrences, it can be a potential natural-derived and low-cost sorbent for the pollution removal in water.

Organic and inorganic matters in soil are both responsible for the sorption of iodine. The iodine content in soil was found to be highly correlated to the amount of oxylate-soluble Al in the near-neutral pH condition and to the amount of oxylate-soluble Fe in the more acidic environment (Whitehead 1978). Generally, I^- and IO_3^- are ionically attracted to the positive charge provided by the free hydroxides of Fe and Al (Whitehead 1984). Table 4 shows iodine adsorption capacity and K_D value of different types of soil and clay. In a batch experiment, the maximum I^- adsorption of 6.59×10^{-3} mg/g occurred in the soil with 4.81% Fe oxide and 5.66% Al oxide, while the maximum IO_3^- adsorbed in the soil with 0.64% Fe oxide and 0.60% Al oxide was 3.41×10^{-3} mg/g (Dai et al. 2009). The experiment was set up with 100 g/L soil dosage at 25 °C for 40 h (Dai et al. 2009).

Nevertheless, in the natural organic constituents of soil, where the positive charge is still scarce, the iodine can still covalently bond and form complexes with functional groups presented in the soil such as phenol, thiol, and amine (Sakuma and Marzukee 1995; Whitehead 1974; Kaplan et al. 2014). The soil samples with higher content of organic matter, organic carbon, calcium carbonate, nitrogen, phosphorus, potassium,

Table 4 Comparison of iodine adsorption in the collected soil, natural clay, and modified clay

Types of soil or clay	Method of modification	[a]Maximum iodine adsorption (mg/g)	Equilibrium time	K_D (mL/g)	Experimental condition					Reference
					Iodine species	Adsorbent dosage (g/L)	Temperature (°C)	pH		
Orthic aridisols	Not modified, collected from Xinjiang, China	3.41×10^{-2}	–	57.60	$IO_3{}^-$	100	25	–		Dai et al. (2009)
Hydragric anthrosols	Not modified, collected from Guangdong, China	6.59×10^{-2}	–	1.97	I^-	100	25	–		Dai et al. (2009)
Sandy loam	Not modified, collected from Zadenhan Iran	–	–	~200	I^-	55.56	9	–		Hosseini et al. (2013)
Entisol	Not modified, collected from Varanasi, India	1.60×10^{-4}	5 h	2.50×10^{-3}	I^-	100	Room	–		Nath et al. (2010)
Argillite	Not modified	–	48 h	0.21	I^-	200	25	–		Tournassat et al. (2007)

(continued)

Table 4 (continued)

Types of soil or clay	Method of modification	[a]Maximum iodine adsorption (mg/g)	Equilibrium time	K_D (mL/g)	Experimental condition				Reference
					Iodine species	Adsorbent dosage (g/L)	Temperature (°C)	pH	
Kaolinite	Not modified	–	1 h	47.09	I^-	20	Room	3.5	Sakuma and Marzukee (1995)
HDPy-smectite	Treated with $C_{21}H_{38}N^+$ Cl^-	–	–	297	I^-	55.56	20	–	Riebe et al. (2005)
HDPy-vermiculite	Treated with $C_{21}H_{38}N^+$ Cl^-	–	–	372	I^-	55.56	20	–	Riebe et al. (2005)
HDPy-bentonite	Treated with $C_{21}H_{38}N^+$ Cl^-	–	–	2800	I^-	200	–	–	Riebe et al. (2001)
HDSMB-bentonite	Treated with $C_{19}H_{42}N^+$ Br^-	–	10 min	3600	I^-	20	25	7	Choung et al. (2014)
ClayFloc™ 750	Commercial bentonite containing quaternary amine	–	7 day	9610	I^-	10	22	5.5	Li et al. (2014), Kaplan et al. (2015)

[a]The value is obtained from the experimental data and not from the mathematical model

and sulfur were found to adsorb iodine (Fukui et al. 1996; Nath et al. 2010). A batch experiment revealed the ability of the collected sandy loam soils to adsorb 1.3×10^{-4} to 1.60×10^{-4} mg/g of I^- adsorption at its capacity (Nath et al. 2010).

Nearly all the sorbate can be desorbed and recovered from the soil (Nath et al. 2010). However, this means that iodine sorption in the soil can be unstable under the condition where pH and temperature fluctuate. The adsorption of iodine may increase and reach its maximum at about 70 °C (Fukui et al. 1996; Hosseini et al. 2013). However, after that point, the adsorption was observed to decrease with the temperature. Higher pH condition decreased the sorption of iodine species in soil due to the OH^- presence, which competes for the adsorption sites (Fukui et al. 1996). The sorption of iodine is affected by the presence of other anions only when the pH value is below neutral (Fukui et al. 1996).

The increasing uptake of I^- and IO_3^- is dependent and independent on the time of contact, respectively (Fukui et al. 1996). Smaller soil grain size promotes a better iodine adsorption due to the increase in surface area (Fukui et al. 1996). However, unlike the activated carbon, IO_3^- is selected for soil instead of iodide. The adsorption of iodate is found to be higher than that of iodide in different temperature and pH condition (Dai et al. 2009; Fukui et al. 1996).

Besides natural soil, clay is another potential adsorbent that has been studied for the removal of iodine. Naturally, clay mineral is limited in its iodine adsorption ability but due to its high retention of cationic species. The non-modified clay, argillite and kaolinite, had K_D value of 0.21 mL/g and 47.09 mL/g, respectively. However, by modifying the materials, iodine removal efficiency could be improved significantly (Kaufhold et al. 2007). By placing inorganic interlayer cations into natural clay, organoclay was produced with available adsorption sites for the iodine (Riebe et al. 2005). For instance, smectite, vermiculite, and bentonite treated with hexadecylpyridinium chloride ($C_{21}H_{38}N^+$ Cl^-) yielded K_D values of 297, 372 and 2800 mL/g, respectively (Riebe et al. 2001, 2005). More efficiency of 3600 mL/g was observed with the K_D value of 3600 mL/g after treating bentonite with hexadecyltrimethylammonium bromide ($C_{19}H_{42}N^+$ Br^-). The commercial clay, ClayFloc™ 750, has the highest K_D value of 9610 mL/g (Li et al. 2014). The adsorbent was designed to contain cage-like structure aluminosilicate minerals, sulfur, iron, nitrogen-containing substitutions as well as quaternary amine functional group (Kaplan et al. 2015).

Although most of the study regarding the iodine adsorption of soil and clay presented the efficiency in the form of coefficient, K_D, overall, the collected soil was found to have very low maximum iodine adsorption capacity. Nevertheless, like the activated carbon, modification of clay was found to have higher efficiency than that of the unmodified adsorbent. In this category, the commercial product, ClayFloc™ 750, exhibited the highest K_D value of 9610 mL/g. HDPy-bentonite and HDSMB-bentonite had over 2000 mL/g K_D value. The unmodified entisol and argillite were presented with the lowest value of K_D.

3.3 Mineral

Since it could be found in soil and clay, the inorganic matter or minerals adsorb iodine ions in various ways. IO_3^- and I^- are adsorbed to the mineral with the mechanism of specific adsorption (replacement of hydroxyl group) and electrostatic force, respectively (Evans and Hammad 1995). Minerals containing metal ions adsorb iodine and form insoluble metal iodides, while anionic exchange process is also possible for those that contain other anions beside the ions of iodine (Sazarashi et al. 1994). Several researchers have studied the iodine sorption ability of different kinds of inorganic matters. Table 5 shows the summary of iodide sorption capacity on minerals and rocks from various studies.

Zeolite is a group of hydrated aluminosilicate mineral. The examples of natural zeolite include clinoptilolite, mordenite, and chabazite. The commercial zeolite was capable of removing a high amount of positive radioactive ions of strontium, barium, and cesium (Sato et al. 2011; Lonin et al. 2015). However, it could absorb a minimal amount of iodine. With the initial iodine concentration of 40–50 μg/L, the commercial zeolite removed up to about 20% of I^- and less than 10% of IO_3^- (Sato et al. 2011). In a dynamic adsorption test, with a 4-day contact time, 0.1 g natural zeolite yielded the maximum K_D value for 1.65×10^5, which seems to be a huge number of the coefficient (Lonin et al. 2015). However, in the same research, the K_D values of zeolite for Nd(III) and Sm(III) were observed to be more than three times that of iodine at 7.48×10^5 and 5.47×10^6, respectively (Lonin et al. 2015).

A modified form of zeolite might be more promising for the removal of iodine in water. Silver ion-exchange zeolite has been observed to remove iodine at the capacity of approximately 1.5 mol/g or 190.356 g/g in the air stream (Choi et al. 2001). Modifying zeolite to include hydroxyapatite and hydroxyfluorapatite increased the stability of iodine captured by the sorbent from the air stream (Watanabe et al. 2009). Although methodologies in the mentioned literature are for the removal of radioactive in the gas phase, the same techniques and sorbent can be adapted for the removal contaminant from the aqueous phase, in the future.

Cinnabar exhibited high K_D value as well as removal percentage (Sazarashi et al. 1994). The hydrotalcite also had a relatively high iodine adsorption capacity of 192.5 mg/g (Paredes et al. 2006). When the adsorbent was doped with silver, to produce a silver-hydrotalcite, the iodine uptake capacity increased to 495.5 mg/g (Bo et al. 2016). On the other hand, at optimum conditions, allophane, attapulgite, and chalcopyrite exhibited low K_D value and removal percentage (Sazarashi et al. 1994). For bismuth oxide, with 20 mg/L initial concentration, the adsorbent took 10 h to remove 70 and 80% of I^- and IO_3^-, respectively (Liu et al. 2016). The adsorbent may remove more IO_3^- than I^-, but the K_D values suggested that the adsorbent favor the removal of the latter (Liu et al. 2016).

The bismuth oxide was very stable in water at pH 7 as only 2.3% of iodine leaching was reported after a 12-h leaching experiment (Liu et al. 2016). However, as the pH increased, the sorption capacity decreased (Liu et al. 2016). With the anion competition experiment, the bismuth oxide has shown a strong I^- selectivity in the

Table 5 Comparison of iodine adsorption on various kinds of minerals

Mineral	[a]Maximum capacity (mg/g)	Removal (%)	K_D (mL/g)	Equilibrium time	Experimental condition				Reference
					Iodine species	Initial iodine concentration (mg/L)	Adsorbent dosage (g/L)	pH	
Zeolite	–	~20	–	30 min	I⁻	4×10^{-3} to 50×10^{-3}	–	7	Sato et al. (2011)
Chalcocite	0.713	56	1285	30 min	I⁻	1	1	5.81	Anderson et al. (1995)
Cuprite	N/A	75	1.5	–	I⁻	12.6	50	7.7	Lefèvre et al. (2000)
Argillite	N/A	N/A	1.7	14 day	I⁻	3.78×10^{-7} - 1260	4	12.6	Devivier et al. (2004)
Cuprous sulfide	60.91	86	370	10 day	I⁻	–	20	8.5–9	Lefèvre et al. (2003)
Allophane	–	13.3	3.07	–	I⁻	1.26×10^{-4}	50	6.3	Sazarashi et al. (1994)
Attapulgite	–	4.57	0.958	–	I⁻	1.26×10^{-4}	50	6.3	Sazarashi et al. (1994)
chalcopyrite	–	43.3	15.1	–	I⁻	1.26×10^{-4}	50	6.3	Sazarashi et al. (1994)
Cinnabar	–	99.9	2×10^4	–	I⁻	1.26×10^{-4}	50	6.3	Sazarashi et al. (1994)
Hydrotalcites	192.5	95.10	–	1 day	I⁻	1.68×10^4	10	–	Paredes et al. (2006)
Silver-hydrotalcites	495.5	>90	–	–	I⁻	500	2	–	Bo et al. (2016)

(continued)

Table 5 (continued)

Mineral	[a]Maximum capacity (mg/g)	Removal (%)	K_D (mL/g)	Equilibrium time	Experimental condition				Reference
					Iodine species	Initial iodine concentration (mg/L)	Adsorbent dosage (g/L)	pH	
Bismuth oxide	~70	–	1.85×10^4	10 h	I^-	20	0.2	7	Liu et al. (2016)
Bismuth oxide	~80	–	1.34×10^4	10 h	IO_3^-	20	0.2	7	Liu et al. (2016)
Silica coated with magnetite and imidazole	130	98	–	40 min	I^-	254	0.45	7	Madrakian et al. (2012)

[a]The value is obtained from the experimental data and not from the mathematical model

presence of other anions, but CO_3^{2-} or Cl^- could pose a strong competition for the sorption sites (Liu et al. 2016). Lastly, the researchers studied the removal of iodine in the synthetic seawater, using Fukushima Nuclear Power Plant as a reference. The bismuth oxide could remove 60–100% of the iodine in the seawater, depending on the initial concentration of iodine added (Liu et al. 2016).

The metallic mineral could be modified for iodine adsorption. The silica was coated with magnetite and imidazole. The sorbent was brought for the iodine removal, and the result was satisfactory with 98% iodine removal and about 130 mg/g maximum final sorption capacity (Madrakian et al. 2012). Up to 90% of iodine was recovered from the sorbent (Madrakian et al. 2012). ZIF 8 or ($Zn(2\text{-methyl-imidazolate})_2$) a type of metal–organic framework (MOF) which contained zinc was studied for its ability to remove iodine vapor (Hughes et al. 2013). A gram of ZIF 8 adsorbed up to 0.66 g of iodine (Hughes et al. 2013). The research on this sorbent should be continued in the future with the sorption experiment in the water.

A natural material called gossypol was modified to have zeolite-like morphology. The material is called gossypol P3 polymorph and is described as a phenolic aldehyde pigment, extracted from cotton plants and seeds. It was investigated for its ability to remove iodine in the aqueous solution. The results confirmed for an iodine inclusion, which the molecules enter the extended cavities of the channels within the zeolite-like structure of the gossypol P3 polymorph (Talipov et al. 2007). The sorbent should be subjected to a batch experiment in the future.

Several kinds of minerals have shown promising results. The silver-hydrotalcite had almost 500 mg/g iodide removal capacity and has been able to remove I^- in the seawater, so it becomes one of the best adsorbents in this category. The results of bismuth oxide also exhibited high K_D for both I^- and IO_3^-, and it is one of a few adsorbents that have been reported for the successful removal of IO_3^-. Consequently, the mineral is considered as one of the most essential adsorbents that contribute to the radioactive iodine removal and should be studied further for better results.

3.4 Polysaccharide

Starch and iodine have been known to react to form unique complex formation. The reaction has often been used in the analytical chemistry field for over a century (Szente et al. 1999). It was claimed that iodine form a complex with both amylose and amylopectin in the presence of water (Saibene and Seetharaman 2006). As a result, iodine can be used to determine amylose and amylopectin content in various kinds of starches such as those in corns and potatoes (Rendleman 2003). Knowing that such complex formation exists, there were some attempts to use natural and synthetic polysaccharide in iodine sorption (Table 6).

Chitin is a natural polysaccharide which was found in 1884. It is the most abundant natural polysaccharide, second to cellulose, found in the exoskeleton of arthropods like shrimps and crabs. Chitin and its ability to bind iodine have been studied for drug delivery (Takahashi 1987). Chitin was reacted with iodine, and the results showed that

Table 6 Iodine adsorption by polysaccharides

Polysaccharide	[a]Maximum capacity (mg/g)	Equilibrium time	Experimental condition					Reference
			Iodine species	Initial iodine concentration (mg/L)	Adsorbent dosage (g/L)	Temperature	pH	
Chitin	116.67	24 h	I_2	$\sim 3.8 \times 10^4$	100	–	–	Takahashi (1987)
Chitosan	100	24 h	I_2	$\sim 5 \times 10^4$	2	25	–	Chen and Wang (2001)
Cyclodextrin–chitosan complex	1200	24 h	I_2	$\sim 5 \times 10^4$	2	25	–	Chen and Wang (2001)
Calcium alginate–silver chloride	152.032	–	I^-	5.5–10	5	25–55	4–10	Zhang et al. (2011)
TiO_2-Fe_2O_3-PVA alginate beads	~ 20	3 h	I^-	200	50	25	8	Majidnia and Idris (2015)

[a]The value is obtained from the experimental data and not from the mathematical model

1 and 3 g of chitin were bound to 0.149 and 0.35 g of iodine, respectively (Takahashi 1987). A report presented the synthesis of chitin aerogel with silver impregnation, which the porous adsorbent was capable of adsorbing I_2 from the vapor (Gao et al. 2017).

Chitosan is another kind of polysaccharide, which was made from chitin by the process of alkaline deacetylation. It is considered to be one of a few natural cationic polymers. It is soluble in aqueous solution and can be modified into many forms such as gel, films, and fiber. In a batch experiment using 0.1 g chitosan beads in 50-mL ethanol solution of various iodine concentrations, the maximum sorption capacity was found to be about 0.2 g iodine per a gram of chitosan (Chen and Wang 2001). A K_D value of 88 mL/g was presented in another study (Li et al. 2014).

Cyclodextrin is a sugar molecule that linked to form a ring. It is produced from starch, and like chitin and chitosan, it has been studied extensively for food and drug delivery system (Chen and Wang 2001). A study exhibited a process of cyclodextrin entrapping the gaseous radioactive iodine in a nuclear waste management (Szente et al. 1999). Although the study was done for the air contamination, it shows that cyclodextrin could adsorb iodine in the presence of water vapor. A complex of cyclodextrin–chitosan was found to have a maximum iodine removal capacity of 1200 mg/g (Chen and Wang 2001).

The alginate is another adsorbent that has been researched for the iodine adsorption. It has been proposed that mucilaginous layers, the cells that cover the surface of brown algae, play an important part of iodine sorption ability (Loderio et al. 2005). Within these mucilaginous layers, alginic acid is contained, which is the main ingredient of sodium alginate. Sodium alginate is a compound that is used for pharmaceutical and commercial chelator product to pull I-131 and Sr-90 from bodies of the consumers (Loderio et al. 2005; Sutton et al. 1971). However, the alginate is an anionic polysaccharide. As a result, some studies impregnated it with various kinds of metals, for iodine sorption in the batch experiments. The iodine adsorption on calcium alginate–silver chloride composite yielded the adsorption capacity within the range of 132.742–152.032 mg/g across a set of initial iodine concentration and temperature (Zhang et al. 2011).

The alginate could also be taken to react with Polyvinyl alcohol (PVA), titania (TiO_2), and maghemite (γ-Fe_2O_3) to create a unique sorbent (Majidnia and Idris 2015). The TiO_2-Fe_2O_3-PVA alginate beads were shown to adsorb iodine in an aqueous solution with about 20 mg/g maximum uptake (Majidnia and Idris 2015). pH 8 provided the optimum condition for the sorption mechanism as the highest iodine percentage removal of about 98%, in 200 mg/L iodide initial concentration, was obtained (Majidnia and Idris 2015). Additionally, it was shown that TiO_2-Fe_2O_3-PVA alginate beads can be reused for more than seven times with only 10% reduction of sorption efficiency is observed at the seventh usage (Majidnia and Idris 2015).

Compared to those of the other categories, the research on polysaccharide was found to be limited. Still, the cyclodextrin–chitosan complex exhibited one of the most iodine uptake capacities across all the adsorbent categories. The complex was only studied for the binding of I_2, but, like the calcium alginate–silver chloride, it can possibly be modified for the removal of the anionic species in the natural environment. Since polysaccharide is widely available, further research should focus on identifying the potential adsorbents.

4 Comparative Study of Natural Adsorbents to the Conventional Anion-Exchange Resin

The ion-exchange resin is the adsorbent often used in conventional water softening and treatment. It is also a useful material for several other practices including medical procedure, food production, and agriculture. Ion-exchange resins usually contain a cross-linked polymer matrix with ion-active sites throughout the structure. It has an ability to adsorb ions with a minimal effect on the chemical formation of the solutes and the form of targeted ion (DOWEX 2000). Ions can be recovered from the sorbent in the process called regeneration, which returns the resin to its initial state.

The anion-exchange resin is usually produced from polystyrene. It is engineered to have quaternary amine on the surface. Attached to the N^+ is Cl^- or OH^- that can be exchanged with the target anionic contaminant. Several commercial anion-exchange resins were used to remove I-129 from the underground water environment. The commercial resin has a total iodine maximum loading capacity of 8.27 $\mu g/g$ (Parker et al. 2014). I^- is observed to be a more desirable species for the sorbent with the maximum capacity of 5.90 $\mu g/g$ (Parker et al. 2014). On the other hand, only about less percentage removal of IO_3^- was reported without the value of maximum capacity (Parker et al. 2014). A recent study on the iodide adsorption by silica-based poly(4-vinyl pyridine) resin yielded the contaminant removal capacity of 148.23 mg/g (Ye et al. 2019).

It would be inaccurate to directly compare the adsorption capacity and equilibrium time of different adsorbent to the anion-exchange resin. This is because the experiments were performed in different condition with different iodine quantification techniques. However, the summary of the sorption mechanism, maximum capacity or K_D, and prices of natural adsorbents and anion-exchange resin are shown in Table 7. The sorption mechanism is different for all the natural adsorbent. The price for activated carbon is cheap but the adsorbent itself relies on the modification to improve its iodine adsorption capacity. The commercial bentonite, ClayFloc™ 750, may also have one of the highest K_D values, but it requires the process of synthesis with organic compound containing quaternary amine. The amylose and amylopectin in polysaccharide are also responsible for the binding of I_2 but like the activated carbon, in order to improve the adsorption capacity of anionic iodine species, the adsorbent still depends extensively on the modification with the organic and inorganic with cationic functional groups. The cyclodextrin–chitosan complex seems to have the highest maximum capacity. Nevertheless, only I_2 was used in the experiment, and the adsorbent is expensive compared to the others.

Table 7 Comparison between the natural adsorbent and conventional anion-exchange resin

Types of natural adsorbent	Adsorption mechanism	Maximum I^- adsorption capacity or K_D from a specific adsorbent	Price/g for specific adsorbent
polystyrene-based Anion-exchange resin	Ion-exchange Cl^- or OH^- with the anionic species of I^-	5.90 µg/g in the underground water environment	~0.84 USD for Dowex-1 chloride form[a]
Activated carbon	1. London dispersion forces 2. Forming bonds with functional group carbonyl, phenol, lactone and carboxyl group 3. Ionic bonding, if modified with metal	305 mg/g from carbon fiber 126.9 mg/g from silver-impregnated carbon	<0.01 USD for untreated, granular, 6 × 12 mesh[b]
Soil and clay	1. Covalently bonded and form complexes with functional groups such as phenol, thiol, and amine 2. Ionically and electrostatically attracted to the positive charge provided by the free hydroxides of Fe and Al	9610 mL/g from the commercial bentonite containing quaternary amine	–
Mineral	1. Specific adsorption (replacement of hydroxyl group) 2. Electrostatic force to cationic components 3. Ion-exchange at the quaternary ammonium, if modified with quaternized agents	192.5 mg/g from hydrotalcites 495.5 mg/g from silver-hydrotalcites 70–80 mg/g from bismuth 130 mg/g silica coated with magnetite and imidazole	<0.01 USD for 4A zeolite[c] ~0.11 USD for hydrotalcite[a] ~0.07 USD for bentonite[a] ~4.58 USD for bismuth (III) oxide[a]
Polysaccharide	1. Form complex with the polysaccharide functional groups; amylose and amylopectin 2. Physical precipitation, ionic bonding, and electrostatic force, if modified with metal	1200 mg/g from cyclodextrin–chitosan complex 152.032 mg/g from calcium alginate–silver chloride	~3.11 USD for β-Cyclodextrin[a]

[a]https://www.sigmaaldrich.com (accessed on 10 January 2019)
[b]https://www.kemcore.com (accessed on 12 January 2019)
[c]https://chemlab.en.alibaba.com/?spm=a2700.details.cordpanyb.1.2f07404cJVc1mg (accessed on 12 January 2019)

5   Conclusions and Future Work

From the literature compiled in this report, the problem of radioactive iodine in the water is not a new phenomenon. The research on iodine adsorption has been studied extensively to remove the contaminant from the water so that human and other living thing are not at risk. The natural adsorbents are abundant and can be easily obtained. Some of the natural adsorbents seem to outperform the conventional anion-exchange. With 1200 mg/g iodine adsorption capacity, cyclodextrin–chitosan complex seems to be one of the promising adsorbents, but it is still only efficient for removing I_2 and quite expensive compared to the other adsorbents. On the other hand, the adsorbent like activated carbon, clay, mineral, and alginate rely on the modification to yield a high adsorption efficiency. 495.5 mg/g I^- removal was reported for the silver-hydrotalcites. The carbon fiber and calcium alginate–silver chloride adsorb up to 305 mg/g and 152.032 mg/g of I^-, respectively. The bentonite, which was synthesized to contain quaternary amine, has a very high K_D value of 9610 mL/g.

Most of the adsorbent seem to exhibit the selectivity toward I^-. Only the collected clay and bismuth oxide were reported to adsorb more IO_3^- than I^-. The activated carbon was found to adsorb I^- and I_2, while the chitin, chitosan, and cyclodextrin, unless modified with cationic-containing agent, have only been reported for I_2 inclusion. Moving forward, the batch experiment of the potential adsorbent should be performed with both IO_3^- than I^- as these are prominent species that exist in the water.

In addition, it is recommended that the kinetic and batch experiment are conducted on the potential materials like gossypol, chitin aerogel, and ZIF 8. These adsorbents were successfully tested for the adsorption of gaseous iodine, so they might be able to adsorb the iodine as well in the water. In term of modification, there are still more room for impregnation and synthesis so that the natural materials contain the desired functional group. Besides, the research on using plant materials as iodine adsorbent is still limited. The experiments should be conducted on plant parts that contain the cationic components so that we may uncover new kinds of natural adsorbent.

Lastly, if these materials are found to be highly selective for the iodine contaminants, we can benefit from them not only in term of environmental safety, but also economical values and industrial applications. Continuing the research on this topic is crucial in leading us to the more sustainable future and creating us a better well-being of both human and other living creatures.

References

Adams, R. E., & Browning, W. E., Jr. (1960). Removal of radioiodine from air streams by activated charcoal.

Agency for Toxic Substances and Disease Registry. (2004). *Toxicological profile for iodine*. US Department of Health and Human Services.

Aldahan, A., Kekli, A., & Possnert, G. (2006). Distribution and sources of 129I in rivers of the Baltic region. *Journal of Environmental Radioactivity, 88,* 49–73.

Anderson, H. L., Balsley, S. D., & Brady, P. V. (1995). Iodide retention by cinnabar (HgS) and Chalcocite (Cu_2S)

Barkauskas, J., & Dervinyte, M. (2004). An investigation of the functional groups on the surface of activated carbons. *Journal of the Serbian Chemical Society, 69,* 363–375.

Bo, A., Sarina, S., Liu, H., Zheng, Z., Xiao, Q., Gu, Y., et al. (2016). Efficient removal of cationic and anionic radioactive pollutants from water using hydrotalcite-based getters. *ACS Applied Materials & Interfaces, 8,* 16503–16510.

Boonamnuayvitaya, V., Chaiya, C., Tanthapanichakoon, W., & Jarudilokkul, S. (2004). Removal of heavy metals by adsorbent prepared from pyrolyzed coffee residues and clay. *Separation and Purification Technology, 35,* 11–22.

Boonamnuayvitaya, V., Sae-Ung, S., & Tanthapanichakoon, W. (2005). Preparation of activated carbons from coffee residue for the adsorption of formaldehyde. *Separation and Purification Technology, 42,* 159–168.

Braverman, L. E., & Cooper, D. (2005). *The thyroid. A fundamental and clinical text* (7th ed.). Lippincott Williams & Wilking.

Chae, S., & Kin, C.-H. (2014). Removal of cesium (137 Cs) and iodide (127 I) by microfiltration nanofiltration reverse osmosis membranes. *Journal of the Korean Society of Water and Wastewater, 28,* 549–554.

Chen, S., & Wang, Y. (2001). Study on β-cyclodextrin grafting with chitosan and slow release of its inclusion complex with radioactive iodine. *Journal of Applied Polymer Science, 82,* 2414–2421.

Choi, B. S., Park, G. I., Kim, J. H., Lee, J. W., & Ryu, S. K. (2001). Adsorption equilibrium and dynamics of methyl iodide in a silver ion-exchanged zeolite column at high temperatures. *Adsorption 7,* 91–103.

Choung, S., Kim, M., Yang, J. S., Kim, M. G., & Um, W. (2014). Effects of radiation and temperature on iodide sorption by surfactant-modified bentonite. *Environmental Science and Technology, 48,* 9684–9691.

DOWEX. (2000). Dowex ion exchange resins: Fundamentals of ion exchange.

Dai, J. L., Zhang, M., Hu, Q. H., Huang, Y. Z., Wang, R. Q., & Zhu, Y. G. (2009). Adsorption and desorption of iodine by various chinese soils: II iodide and iodate. *Geoderma, 153,* 130–135.

Delle Site, A. (2001). Factors affecting sorption of organic compounds in natural sorbent/water systems and sorption coefficients for selected pollutants. A review. *Journal of Physical and Chemical Reference Data, 30,* 187–439.

Denham, M., Kaplan, D., & Yeager, C. (2009). Groundwater radioiodine: Prevalence, biogeochemistry, and potential remedial approaches. Savannah River National Laboratory.

Devivier, K., Devol-Brown, I., & Savoye, S. (2004). Study of iodide sorption to the argillite of Tournemire in alkaline media. *Applied Clay Science, 26,* 171–179.

Eden, G. E., Downing, A. L., & Wheatland, A. B. (1952). Observations on the removal of radioisotopes during purification of domestic water supplies: Radio-iodine. *Journal of the Institution of Engineers, 6,* 511–532.

Evans, G. J., Hammad, K. A. (1995). Radioanalytical studies of iodine behavior in the environment. *Journal of Radioanalytical and Nuclear Chemistry, 192.*

Fukuda, S., Iwamoto, K., Atsumi, M., Yokoyama, A., Nakayama, T., Ishida, K., et al. (2014). Global searches for microalgae and aquatic plants that can eliminate radioactive cesium, iodine and strontium from the radio-polluted aquatic environment: A bioremediation strategy. *Journal of Plant Research, 127,* 79–89.

Fukui, M., Fujikawa, Y., & Satta, N. (1996). Factors affecting interaction of radioiodide and iodate species with soil. *Journal of Environmental Radioactivity, 31,* 199–216.

Gao, R., Lu, Y., Xiao, S., & Li, J. (2017). Facile fabrication of nanofibrillated Chitin/Ag_2O heterostructured aerogels with high iodine capture efficiency. *Scientific Reports, 7,* 1–9.

Hamasaki, T., Nakamichi, N., Teruya, K., & Shirahata, S. (2014). Removal efficiency of radioactive cesium and iodine ions by a flow-type apparatus designed for electrochemically reduced water production. *PLoS ONE 9.*

Hoskins, J. S., & Karafil, T. (2002). Removal and sequestration of iodide using silver-impregnated activated carbon. *Environmental Science and Technology, 36,* 784–789.

Hosseini, S. A., Hosseini, S. A., & Mehr, J. P. J. (2013). The role of temperature on radio iodine adsorption behavior in the sandy-loam soil. *Asian Journal of Scientific Research, 6,* 129–134.

Hosseini, S. A. (2010). The effect of marine algae on radioactivity adsorption of iodine in drinking water. *Asian Journal of Applied Sciences,* 1–5.

Hou, X., Povinec, P. P., Zhang, L., Shi, K., Biddulph, D., Chang, C., et al. (2013). Iodine-129 in seawater offshore Fukushima: Distribution, inorganic speciation, sources, and budget. *Environmental Science & Technology, 47,* 3091–3098.

Hu, Q. H., Weng, J. Q., & Wang, J. S. (2010). Sources of anthropogenic radionuclides in the environment: A review. *Journal of Environmental Radioactivity, 101,* 426–437.

Hughes, J. T., Sava, D. F., Nenoff, T. M., & Navrotsky, A. (2013). Thermochemical evidence for strong iodine chemisorption by ZIF-8. *Journal of the American Chemical Society, 135,* 16256–16259.

Jhiang, S. M., Smanik, P. A., & Mazzaferri, E. L. (1994). Development of a single-step duplex RT-PCR detecting different forms of ret activation, and identification of the third form of in vivo ret activation in human papillary thyroid carcinoma. *Cancer Letters, 78,* 69–76.

Kamei, D., Kuno, T., Sato, S., Nitta, K., & Akiba, T. (2012). Impact of the Fukushima Daiichi nuclear power plant accident on hemodialysis facilities: An evaluation of radioactive contaminants in water used for hemodialysis. *Therapeutic Apheresis and Dialysis, 16,* 87–90.

Kaplan, D. I., Denham, M. E., Zhang, S., Yeager, C., Xu, C., Schwehr, K. A., et al. (2014b). Radioiodine biogeochemistry and prevalence in groundwater. *Critical Reviews in Environment Science and Technology, 44,* 2287–2335.

Kaplan, D. I., Zhang, S., Roberts, K. A., Schwehr, K., Xu, C., Creeley, D., et al. (2014). Radioiodine concentrated in a wetland. *Journal of Environmental Radioactivity, 131.*

Kaplan, D. I., Knox, A. S., Crapse, K. P., Li, D., & Diprete, P. (2015). Organo-modified clays for Removal of aqueous radioactive anions.

Kaufhold, S., Pohlmann-Lortz, M., Dohrmann, R., & Nüesch, R. (2007). About the possible upgrade of bentonite with respect to iodide retention capacity. *Applied Clay Science, 35,* 39–46.

Kawamura, H., Kobayashi, T., Furuno, A., In, T., Ishikawa, Y., Nakayama, T., Shima Sm Awaji, T. (2011). Preliminary numerical experiments on oceanic dispersion of 131I and 137Cs discharged into the ocean because of the Fukushima Daiichi nuclear power plant. *Journal of Nuclear Science and Technology, 48,* 1349–1356.

Keogh, S. M., Aldahan, A., Possnert, G., Finegan, P., Leon Vintro, L., & Mitchell, P. I. (2007). Trends in the spatial and temporal distribution of 129I and 99Tc in coastal waters surrounding Ireland using *Fucus vesiculosus* as a bio-indicator. *Journal of Environmental Radioactivity, 95,* 23–38.

Kitada, S., Oikawa, T., Watanabe, S., Nagai, K., Kobayashi, Y., Matsuki, M., et al. (2015). Removal of radioactive iodine and cesium in water purification. *Desalination and Water Treatment, 54,* 3494–3501.

Kohman, T. P., & Edwards, R. R. (1966). Iodine-129 as a geochemical and ecological tracer, progress report to Environmental Sciences Branch.

Kosaka, K., Asami, M., Kobashigawa, N., Ohkubo, K., Terada, H., Kishida, N., et al. (2012). Removal of radioactive iodine and cesium in water purification processes after an explosion at a nuclear power plant due to the Great East Japan Earthquake. *Water Research, 46,* 4397–4404.

Küpper, F. C., Schweigert, N., Ar Gall, E., Legendre, J. M., Vilter, H., & Kloareg, B. (1998). Iodine uptake in Laminariales involves extracellular, haloperoxidase-mediated oxidation of iodide. *Planta, 207,* 163–171.

Lefevre, G., Walcarius, A., Ehrhardt, J. J., & Bessiere, J. (2000). Sorption of iodide on cuprite (Cu_2O). *Langmuir, 16,* 4519–4527.

Lefèvre, G., Bessière, J., Ehrhardt, J. J., & Walcarius, A. (2003). Immobilization of iodide on copper(I) sulfide minerals. *Journal of Environmental Radioactivity, 70,* 73–83.

Li, D., Kaplan, D. I., Knox, A. S., Crapse, K. P., & Diprete, D. P. (2014). Aqueous 99Tc, 129I and 137Cs removal from contaminated groundwater and sediments using highly effective low-cost sorbents. *Journal of Environmental Radioactivity, 136,* 56–63.

Liu, S., Kang, S., Wang, H., Wang, G., Zhao, H., & Cai, W. (2016). Nanosheets-built flowerlike micro/nanostructured $Bi_2O_{2.33}$ and its highly efficient iodine removal performances. *Chemical Engineering Journal, 289,* 219–230.

Loderio, P., Cordero, B., Barriada, J. L., Herrero, R., & Sastre de Vicente, M. E. (2005). Biosorption of cadmium by biomass of brown marine macroalage. *Bioresource Technology, 96,* 1796–1803.

Lonin, A. Y., Levenets, V. V., Neklyudov, I. M., & Shchur, A. O. (2015). The usage of zeolites for dynamic sorption of cesium from waste waters of nuclear power plants. *Journal of Radioanalytical and Nuclear Chemistry, 303,* 831–836.

Madrakian, T., Afkhami, A., Zolfigol, M. A., Ahmadi, M., & Koukabi, N. (2012). Application of modified silica coated magnetite nanoparticles for removal of iodine from water samples. *Nano-Micro Letters, 4,* 57–63.

Majidnia, Z., & Idris, A. (2015). Photocatalytic reduction of iodine in radioactive waste water using maghemite and titania nanoparticles in PVA-alginate beads. *Journal of the Taiwan Institute of Chemical Engineers, 54,* 137–144.

Mimides, T., & Kathariou, L. (2005). Radioactive iodine in terrestrial ecosystems (pp. 1010–1017).

Miyake, Y., Matsuzaki, H., Fujiwara, T., Saito, T., Yamagata, T., Honda, M., et al. (2012). Isotopic ratio of radioactive iodine (129I/131I) released from Fukushima Daiichi NPP accident. *Geochemical Journal, 46,* 327–333.

Morita, T., Niwa, K., Fujimoto, K., Kasai, H., Yamada, H., Nishiutch, K., et al. (2010). Detection and activity of iodine-131 in brown algae collected in the Japanese coastal areas. *Science of the Total Environment, 408,* 3443–3447.

Nath, T., Raha, P., & Rakshit, A. (2010). Sorption and desorption behaviour of iodine in alluvial soils of Varanasi. *India. Agricultura, 14,* 9–14.

National Academy of Science. (1999). Exposure of the American people to iodine-131 from Nevada nuclear-bomb tests: Review of the National Cancer Institute report and public health implication.

Nitschke, U., & Stengel, D. B. (2014). Iodine contributes to osmotic acclimatisation in the kelp *Laminaria digitata* (Phaeophyceae). *Planta, 239,* 521–530.

Nwosu, F. O., Olu-Owolabi, B. I., Adebowale, K. O., Henle, T., & Schwarzenbolz, U. (2009). Pore structure and surface functional groups on six tropical fruit nutshell active carbons. *Bioremediation, Biodivers Bioavailab.*

O'Dowd, C. D., Jimenez, J. L., Bahreini, R., Flagan, R. C., Seinfeld, J. H., Hämeri, K., et al. (2002). Marine aerosol formation from biogenic iodine emissions. *Nature, 417,* 632–636.

Paredes, S. P., Fetter, G., Bosch, P., & Bulbulian, S. (2006). Iodine sorption by microwave irradiated hydrotalcites. *Journal of Nuclear Materials, 359,* 155–161.

Parker, K. E., Golovich, E. C., & Wellman, D. M. (2014). *Iodine adsorption on ion-exchange resins and activated carbons—batch testing.* VA: Alexandria.

Peterson, J., MacDonell, M., Haroun, L., & Monette, F. (2007). Radiological and chemical fact sheets to support health risk analyses for contaminated areas. Argonne National Laboratory, 38–9.

Plazinski, W. (2012). Sorption of metal cations by alginate-based biosorbents. On the correct determination of the thermodynamic parameters. *Journal of Colloid and Interface Science, 368,* 547–551.

Prados-Roman, C., Cuevas, C. A., Fernandez, R. P., Kinnison, D. E., Lamarque, J. F., & Saiz-Lopez, A. (2015). A negative feedback between anthropogenic ozone pollution and enhanced ocean emissions of iodine. *Atmospheric Chemistry and Physics, 15,* 2215–2224.

Ramana, M. V. (2009). Nuclear power: Economic, safety, health, and environmental issues of near-term technologies. *Annual Review of Environment and Resources, 34.*

Ravichandran, R., Binukumar, J. P., Sreeram, R., & Arunkumar, L. S. (2011). An overview of radioactive waste disposal procedures of a nuclear medicine department. *Journal of Medical Physics, 36*, 95–99.

Rendleman, J. A. (2003). The reaction of starch with iodine vapor. Determination of iodide-ion content of starch-iodine complexes. *Carbohydrate Polymers, 51*, 191–202.

Riebe, B., Bors, J., & Dultz, S. (2001). Retardation capacity of organophilic bentonite for anionic fission products. *Journal of Contaminant Hydrology, 47*, 255–264.

Riebe, B., Dultz, S., Bunnenberg, C. (2005). Temperature effects on iodine adsorption on organo-clay minerals—I. Influence of pretreatment and absorption temperature. *Applied Clay Science 28*, 9–16.

Rodriguez, N. M., & Marsh, H. (1987). Surface structure of coals studied by iodine and water adsorption. *Fuel, 66*, 1727–1732.

Saibene, D., & Seetharaman, K. (2006). Segmental mobility of polymers in starch granules at low moisture contents. *Carbohydrate Polymers, 64*, 539–547.

Sakuma, S. H., & Marzukee, N. (1995). Removal of iodine-125 from effluents by chemical and soil column methods. *Journal of Radioanalytical and Nuclear Chemistry, 196*, 77–87.

Sancho, M., Arnal, J. M., Verdú, G., Lora, J., & Villaescusa, J. I. (2006). Ultrafiltration and reverse osmosis performance in the treatment of radioimmunoassay liquid wastes. *Desalination, 201*, 207–215.

Sato, I., Kudo, H., & Tsuda, S. (2011). Removal efficiency of water purifier and adsorbent for iodine, cesium, strontium, barium and zirconium in drinking water. *Journal of Toxicological Sciences, 36*, 829–834.

Sazarashi, M., Ikeda, Y., Seki, R., & Youshikawa, H. (1994). Adsorption of I⁻ ions on minerals for ^{129}I waste management. *Journal of Nuclear Science and Technology, 31*, 620–622.

Smith, D. K., Finnegan, D. L., & Bowen, S. M. (2003). An inventory of long-lived radionuclides residual from underground nuclear testing at the Nevada test site, 1951–1992. *Journal of Environmental Radioactivity, 67*, 35–51.

Sun, H., La, P., Yang, R., Zhu, Z., Liang, W., Yang, B., et al. (2017). Innovative nanoporous carbons with ultrahigh uptakes for capture and reversible storage of CO_2 and volatile iodine. *Journal of Hazardous Materials, 321*, 210–217.

Sutton, A., Harrison, G. E., Carr, T. E. F., & Barltrop, D. (1971). Reduction in the absorption of dietary strontium in children by an alginate derivative. *International Journal of Radiation Biology and Related Studies in Physics, Chemistry and Medicine, 19*, 79–85.

Szente, L., Fenyvesi, É., & Szejtli, J. (1999). Entrapment of iodine with cyclodextrins: Potential application of cyclodextrins in nuclear waste management. *Environmental Science and Technology, 33*, 4495–4498.

Sánchez-Polo, M., Rivera-Utrilla, J., Salhi, E., & Gunten, U. (2006). Removal of bromide and iodide anions from drinking water by silver-activated carbon aerogels. *Journal of Colloid and Interface Science, 300*, 437–441.

Takahashi, Y. (1987). Binding properties of alginic acid and chitin. *Journal of Inclusion Phenomena, 5*, 525–534.

Talipov, S. A., Ibragimov, B. T., Tadjimukhamedov, F. K., Tiljakov, Z. G., Blake, A. J., Hertzsch, T., et al. (2007). Inclusion of molecular iodine into channels of the organic zeolite-like gossypol. *Journal of Inclusion Phenomena and Macrocyclic Chemistry, 59*, 287–292.

Tay, T., Ucar, S., & Karagöz, S. (2009). Preparation and characterization of activated carbon from waste biomass. *Journal of Hazardous Materials, 165*, 481–485.

Tournassat, C., Gaucher, E. C., Fattahi, M., & Grambow, B. (2007). On the mobility and potential retention of iodine in the Callovian-Oxfordian formation. *Physics and Chemistry of the Earth, 32*, 539–551.

United Nations Scientific Committee on the Effects of Atomic Radiation. (2000). *Sources and effects of ionizing radiation; Annex J. exposure and effects of Chernobyl accident*. New York.

Van Der Bruggen, B., Vandecasteele, C., Van Gestel, T., Doyen, W., & Leysen, R. (2003). A review of pressure-driven membrane processes in wastewater treatment and drinking water production. *Environmental Progress, 22,* 46–56.

Watanabe, Y., Ikoma, T., Yamada, H., Suetsugu, Y., Komatsu, Y., Stevens, G. W., et al. (2009). Novel long-term immobilization method for radioactive iodine-129 using a zeolite/apatite composite sintered body. *ACS Applied Materials & Interfaces, 1,* 1579–1584.

Whitehead, D. C. (1974). The sorption of iodide by soil components. *Journal of the Science of Food and Agriculture, 25,* 73–79.

Whitehead, D. C. (1978). Iodine in soil profiles in relation to iron and aluminium oxides and Organic Matter. *Journal of Soil Science, 29,* 88–94.

Whitehead, D. C. (1984). The distribution and transformations of iodine in the environment. *Environment International, 10,* 321–339.

Yang, O. B., Kim, J. C., Lee, J. S., & Kim, Y. G. (1993). Use of activated carbon fiber for direct removal of iodine from acetic acid solution. *Industrial and Engineering Chemistry Research, 32,* 1692–1697.

Ye, Z., Chen, L., Liu, C., Ning, S., Wang, X., Wei, Y. (2019) The rapid removal of iodide from aqueous solutions using a silica-based ion-exchange resin. *Reactive and Functional Polymers, 135,* 52-57.

Zhang, H., Gao, X., Guo, T., Li, Q., Liu, H., Ye, X., et al. (2011). Adsorption of iodide ions on a calcium alginate-silver chloride composite adsorbent. *Colloids Surfaces A Physicochem Eng Asp, 386,* 166–171.

Zhang, X., Gu, P., Li, X., & Zhang, G. (2017). Efficient adsorption of radioactive iodide ion from simulated wastewater by nano Cu_2O/Cu modified activated carbon. *Chemical Engineering Journal, 322,* 129–139.

Zhang, S., Xu, C., Creeley, D., Ho, Y. F., Li, H. P., Grandbois, R., et al. (2013). Iodine-129 and iodine-127 speciation in groundwater at the Hanford site, U.S.: Iodate incorporation into calcite. *Environmental Science & Technology, 47,* 9635–9642.

Zhou, J., Hao, S., Gao, L., & Zhang, Y. (2014). Study on adsorption performance of coal based activated carbon to radioactive iodine and stable iodine. *Annals of Nuclear Energy, 72,* 237–241.

An Insight into Characterization, Mechanism and Kinetics of Congo Red Sorption Using Biowaste FRLP

Mohan Rao Tamtam and Basava Rao Vudata Venkata

Abstract Commercial utilization of solid waste is superior to other waste management methods in environment point of view. Dry leaves have potential utility as an adsorbent. The abundantly available biomass *Ficus religiosa* leaf powder (FRLP) was used as an adsorbent for the removal of pollutants like congo red (CR) from aqueous solution. The nature and mechanism for the removal of CR were determined from characterization and kinetic studies. The results reveal the hydrogen bond between various functional groups of adsorbent and the dye. The rate of transport is of pseudo-second-order type.

Keywords *Ficus religiosa* · Congo red · Kinetics · XRD · SEM · FTIR

1 Introduction

Global solid waste generation is around 1.3 billion tonnes per year, with a footprint of 1.2 kg per person per day. By 2025, it is expected to reach 2.2 billion tonnes, with the current phase of population growth, industrialization and urbanization. East Asia and the Pacific region generate approximately 270 million tonnes annually with a per capita of 0.44 kg per person per day. Projected statistics reveals that China and India face serious problems of waste management compared to any other country (http://www.worldbank.org/en/topic/urbandevelopment/brief/solid-waste-management). India is one of the top ten countries producing solid waste with more than 10^5 metric tonnes per day. It is anticipated that population of India would be about 1823 million by 2051 and about 300 million tons per annum of MSW will be generated (http://www.cpcbenvis.nic.in/newsletter/solidwastejun1997/jun1997.htm).

M. R. Tamtam (✉)
Chemical Engineering Department, Bapatla Engineering College, Bapatla, Andhra Pradesh, India

B. R. Vudata Venkata
College of Technology, OU, Hyderabad, Telangana, India

© Springer Nature Singapore Pte Ltd. 2020
S. K. Ghosh et al. (eds.), *Recent Trends in Waste Water Treatment and Water Resource Management*, https://doi.org/10.1007/978-981-15-0706-9_18

Around 60% of solid waste from lower and middle income country like India is organic. And leaf litter is one of the major contributors of organic waste. Dry leaf piles are considered unsightly and shelter for disease-causing agents (http://www.earthamag.org/stories/2018/4/9/dont-burn-those-dry-leaves-put-them-to-good-use-instead). Open air burning is a general practice to alleviate this problem, but it contributes to air pollution, and hence, global warming. Commercial utilization is a superior method over other solid waste management strategies. Recent literature review proves the capability of dry leaves as a potential adsorbent in chemical industrial sector and water treatment plants (Mohan Rao and Basava Rao 2016).

In this present study, *Ficus religiosa* leaves are powdered and used as an adsorbent to recover synthetic dye, Congo red dye from water effluents. This paper focuses on the characterization of adsorbent, the binding mechanism and the kinetics.

2   Materials and Methods

2.1   Preparation of Anionic Dye Solutions

Congo red (CI = 22,120) was supplied by Merck (Mumbai, India). Initially, a solution of 1000 mg/L was prepared and then diluted as per the requirement.

2.2   Preparation of the Adsorbent

Ficus religiosa leaves were collected directly from the trees in the local area, rinsed with water to remove the debris and dried in an oven at 80 °C to a constant weight. The dried leaves were made to powder of required size and finally packed in desiccators.

2.3   Biosorption Experiments

Original dye solution of concentration 1000 mg L^{-1} was prepared. And it was diluted to various concentrations up to 25 mg L^{-1}. Pre-calculated amount of adsorbent (0.01–1 g) was measured accurately with an analytical balance (sX200) and added to 50 mL of feed solution; pH of the solution was measured with a digital pH meter (ELICO-L1 612) and varied (2–10) by using 0.1 N HCl and 0.1 N NaOH solutions. The adsorbent solution was agitated with Remi make Temperature Controlled Orbital Shaker (REMI—CIS 24 BL). At the end, the samples were collected and centrifuged to remove the suspended solid particles using REMI C 24 centrifuge. The clear liquid was collected and analyzed using UV–VIS Spectrophotometer (SYSTRONICS-117) at a wavelength of 498 nm under different conditions.

2.4 Characterization

A scanning electron microscope SEM model EVO 18, Carl Zeiss, was used for the microstructure analysis of the sample surface. EDS (Oxford Instrument) was attached to it for elemental analysis of the sample. The X-ray diffraction of the sample was obtained with SHIMADZU make XRD 7000. Fourier transmission electroscope was used to identify the functional groups on biomass. The sample was examined using SHIMADZU, FTIR 8400S.

3 Results and Discussion

3.1 Characterization

The SEM image at 10,000× magnification is shown in Fig. 1. It shows irregular surface with some small particles.

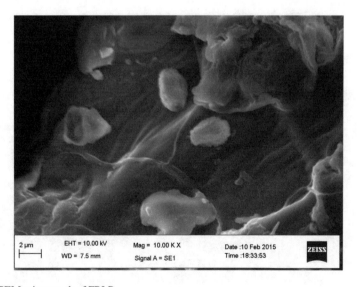

Fig. 1 SEM micrograph of FRLP

X-ray diffraction spectra of FRLP are shown in Fig. 2. The overall XRD spectrum is indicative of an amorphous nature of the adsorbent. Amorphous materials have greater specific surface area and more active sites which render them good adsorbents.

EDS analysis was carried out to identify the minerals present in FRLP. EDS spectrum has been shown in Fig. 3. It shows the presence of minerals like Ca, K, Cu, Na and Mg with C and O as major compositions.

The FTIR spectral data of FRLP are shown in Fig. 4. It shows the presence of several functional groups that can interact with pollutants like CR in the adsorption process. A broad peak in the range of 3300–3450 cm^{-1} is the characteristic

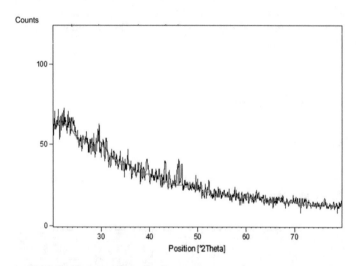

Fig. 2 X-ray diffraction spectra of FRLP

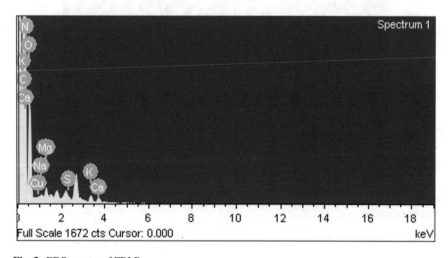

Fig. 3 EDS spectra of FRLP

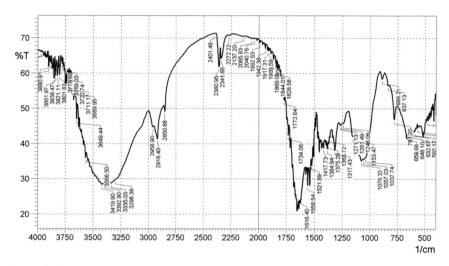

Fig. 4 FTIR spectra of FRLP

of O–H/N–H stretching frequencies and may be attributed to the presence of alcohols, phenols, amines and carboxylic acids in FRLP. The absorbance at 2918 and 2850 cm^{-1} is due to aliphatic C–H stretching in alkanes. While the absorption at 2360 cm^{-1} appears due to C≡C stretching, medium peaks at 2341, 1616 cm^{-1} could be attributed to C–H stretch in cellulose and starch. The weak but sharp peaks at 1558 and 1417 cm^{-1} could be ascribed to C–H deformation in lignin. Similarly, relatively less intense sharp peaks at 1384, 1375 cm^{-1} are characteristics of C–H bending vibrations of alkanes while C–hetero atom stretching frequencies for C–O and C–N bonds at 1076, 1057 cm^{-1} clearly indicates the presence of carboxylic acids/alcohols/amines. The presence of alkyl halides is confirmed by typical C–Cl stretching frequencies at 837 and 781 cm^{-1}, and the absorption for glycosidic linkage in hemi-cellulose appears at 889 cm^{-1}.

Interpretation of analytical data obtained from our studies revealed that FRLP contains various organic components which include alcohols, long-chain hydrocarbons, aliphatic alcohols along with fibers, tannin, different types of proteins and carbohydrates. Besides, it is rich in minerals like calcium, phosphorus, iron, copper, magnesium and sodium. These observations were further supported by the studies conducted by Chitra Gupta (2012).

3.2 Adsorption Kinetics Study

Kinetic studies help to estimate the time of operation and give an idea about the mechanism of the process. In this study, pseudo-first-order and pseudo-second-order kinetic models were applied to determine the suitable rate expression. The linear form of pseudo-first-order rate expression of Lagergren based on solid capacity is generally expressed as follows (Purkait et al. 2007):

$$\ln(q_e - q_t) = \ln q_e - K_1 t \tag{5}$$

where q_e and q_t are uptake at equilibrium and at time t, respectively, and K_1 is the rate constant (min^{-1}). A straight line for the plot of log $(q_e - q_t)$ versus t would suggest the applicability of the model. The values of K_1 and q_e can be calculated from the slope and intercept of this line.

The linear expression of pseudo-second-order kinetic model of Ho and McKay (1999) (Acemioğlu 2004) is given below:

$$\frac{t}{q_t} = \frac{1}{K_2 q_{2e}^2} + \frac{t}{q_{2e}} \tag{6}$$

where K_2 is the rate constant in unit g min^{-1}. A linear relationship to the plot of t/q_t versus t implies applicability of the pseudo-second-order kinetics.

Various experiments were conducted with different initial concentrations (25, 50, 150 mg/L). Samples were withdrawn at different time intervals, and the data were tested for the kinetic models. Kinetic parameters were calculated from the linear plots of pseudo-first- and second-order models (Fig. 5) and the results of which have been given in Table 1. Adequacy of the model was tested in terms of R^2 values and by comparing uptakes that were actual and estimated by the models. The R^2 values for pseudo-first- and second-order models at 25, 50 and 150 mg/L were 0.73, 0.88 and 0.70 and 1, 1 and 0.998, respectively. Estimated equilibrium uptake values of the pseudo-second-order model are very close to that of actual. However, such is not the case of pseudo-first-order. Hence, the adsorption mechanism is closer to a pseudo-second-order process. It is similar to the observations cited for adsorption of Congo red on activated carbon, calcium-rich fly ash and CaCl$_2$ modified bentonite (Purkait et al. 2007; Liping Guo and Wang 2009; Jing et al. 2011).

Fig. 5 Pseudo-second-order kinetics for biosorption of CR using FRLP

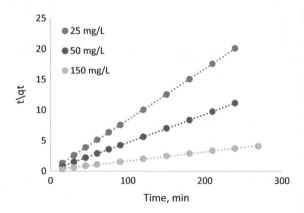

Table 1 Kinetic parameter constants for biosorption of CR using FRLP

C_0 (mg/L)	$q_{e, exp}$ (mg/g)	Pseudo-first-order reaction			Pseudo-second-order reaction		
		K_1 (min^{-1})	q_e (mg/g)	R^2	K_2 (g min^{-1})	q_e (mg/g)	R^2
25	11.94	0.043	2.63	0.73	1748.90	12.05	1
50	21.56	0.026	4.70	0.88	4633.23	21.74	1
150	70.68	0.029	138.24	0.70	26,573.13	71.43	0.998

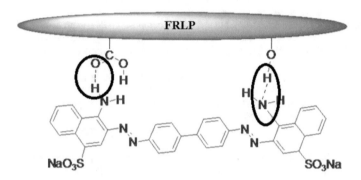

Fig. 6 Proposed mechanistic model for CR adsorption on FRLP

3.3 Mechanistic Studies

Based on the above studies, a type of physical interaction or physisorption model caused by weak hydrogen bond forces of various functional groups present on the surface of FRLP with those of aromatic amine functionalities of Congo red is expected. As could be observed from the proposed mechanistic model depicted in Fig. 6, the basic carbonyl part from the carboxylic acid is H-bonded to the electron-deficient hydrogen atom of the aromatic amine while the simple hydroxyl groups are presumably hydrogen bonded to electron-rich nitrogen atom of the amine. These noncovalent interactions are in agreement with the model proposed earlier (Mao et al. 2008; Kumar et al. 2010) and appear to provide the requisite stabilization for adsorption.

4 Conclusions

FRLP is an amorphous material with greater specific surface area. It contains various organic components like alcohols, hydrocarbons, along with fibers and tannin which are rich in functional groups. Based on the studies, hydrogen bond is expected between carbonyl group of carboxylic acid and the electron-deficient hydrogen atom of the aromatic amine. In addition, simple hydroxyl groups are presumably hydrogen bonded to electron-rich nitrogen atom of the amine. In kinetic studies, pseudo-second-order model best described the sorption kinetics compared to other models.

References

Acemioğlu, B. (2004). Adsorption of Congo red from aqueous solution onto calcium-rich fly ash. *Journal of Colloid and Interface Science, 274*(2), 371–379.

Chitra Gupta, S. S. (2012). Taxonomy, phytochemical composition and pharmacological prospectus of *Ficus religiosa* Linn. (Moraceae)—A review. *The Journal of Hytopharmacology 1*(1), 57–70.

Ho, Y. S., & McKay, G. (1999). Pseudo-second order model for sorption processes. *Process Biochemistry, 34*(5), 451–465.

http://www.cpcbenvis.nic.in/newsletter/solidwastejun1997/jun1997.htm 9/6/2012 11:04:22 AM. Accessed September 2018.

http://www.earthamag.org/stories/2018/4/9/dont-burn-those-dry-leaves-put-them-to-good-use-instead. Accessed September 2018.

http://www.worldbank.org/en/topic/urbandevelopment/brief/solid-waste-management. Accessed September 2018.

Jing, C. X. Y., Jia, Y., Yue, D., Ma, J., & Yin, X. (2011). Adsorption properties of congo red from aqueous solution on modified hectorite: Kinetic and thermodynamic studies. *Desalination, 265,* 81–87.

Kumar, G. V., Ramalingam, P., Kim, M. J., Yoo, C. K., & Kumar, M. D. (2010). Removal of acid dye (violet 54) and adsorption kinetics model of using *Musa* spp. waste: A low-cost natural sorbent material. *Korean Journal of Chemical Engineering, 27,* 1469–1475.

Liping Guo, L. L., & Wang, A. (2009). Use of $CaCl_2$ modified bentonite for removal of Congo red dye from aqueous solutions. *Desalination, 249,* 797–801.

Mao, J., Won, S. W., Min, J., & Yun, Y. S. (2008). Removal of Basic Blue 3 from aqueous solution by *Corynebacterium glutamicum* biomass: Biosorption and precipitation mechanisms, Korean. *Journal of Chemical Engineering, 25,* 1060–1064.

Mohan Rao, T., & Basava Rao, V. V. (2016). Biosorption of Congo Red from aqueous solution by crab shell residue: A comprehensive study. *SpringerPlus 5*(537), 1–14.

Purkait, M. K., Maiti, A., Das Gupta, S., & De, S. (2007). Removal of congo red using activated carbon and its regeneration. *Journal of Hazardous Materials, 145,* 287–295.

Novel Utility of Bacteria of Fish Gut Wastes Collected from Local Markets for Detoxification of Textile Dyes and Recycling of Water for Cultivation

Cherasala Reddi Roja, Vempalli Sumathi, Duvvuri Sai Vinathi, Taticherla Hemalatha, D. C. Reddy and Varada Kalarani

Abstract Preliminary survey shows that an average of 500 kgs of fish waste is being produced per day from each local fish market. Several such markets are in function in whole of Chittoor district alone. These wastes are known to be either underutilized to produce low market value products or dumped in open water systems. Fish waste management has been one of the major problems of concern since it will have the greatest impact on the environment. Fish gut is rich in bacteria; some of which are beneficial or probiotic. Complete utilization of fishery waste for recovering high-end products would be a fruitful strategy to overcome the issue as well to increase the economic gain. Worldwide dyestuff and pigments worth s \$10.6 billion (2008) are being produced. India is the second largest exporter of dyestuffs and intermediates among developing countries, after China. The textile industry accounts for the largest consumption of dyestuffs, at nearly 80%. Dye containing waters are known for reducing efficiency of seed germination, plant growth and elongation of shoot and roots (Nirmalarani and Janardhanan in Madras Agric J 75:41, 1988), carcinogenic and mutagenic (Venturini and Tamaro in Mutation Res 68: 307–312, 1979; Mathur et al. in Appl Ecol Environ 4(1):111–118, 2005), causing skin and eye irritation, reducing aquatic life affecting liver, gill, kidney, intestine and gonads as well as liver damage in humans (Badr et al. in Mol Biol Evol 17:499–510, 2008). In spite of the above, the industrial effluents with greater than admissible levels of dyes are noticed to be released into open water systems. Microorganisms are known to reduce the dyes by secreting enzymes such as laccase, azoreductase, peroxidase and hydrogenase. Reduced forms of dyes are further mineralized into simpler compounds and are utilized as their energy source. Hence in the present study, *Bacillus subtilis* and *Terribacillus saccharophillus,* isolated from the gut of *Labeo rohita* (*Gendi*), the most local delicacy collected from the market, were examined for their ability to decolourize the textile dye, malachite green (MG). UV-Vis studies showed that both *B. subtilis* and *T. saccharophillus* have an ability to degrade methyl red, the azo dye.

C. Reddi Roja · V. Sumathi · D. Sai Vinathi · T. Hemalatha · V. Kalarani (✉)
Department of Biotechnology, Sri Padmavati Mahila Visvavidyalayam (Women's University), Tirupati, India

D. C. Reddy
Former Professor, Department of Fishery Science and Aquaculture, S.V. University, Tirupati, Andhra Pradesh, India

S. K. Ghosh et al. (eds.), *Recent Trends in Waste Water Treatment and Water Resource Management*, https://doi.org/10.1007/978-981-15-0706-9_20

Further, Fourier transform infrared spectroscopy (FTIR) studies of *B. subtilis*- and *T. saccharophillus*-treated dye solutions after 0, 24 and 48 h showed remarkable change and reduction in the number of functional groups with progress in treatment time, clearly illustrating the impact of both these bacteria in degrading the dye molecules. Phytotoxicity studies further clearly demonstrated that addition of *B. subtilis* and *T. saccharophillus* to the growth medium used for seed germination significantly degrades the dye integrity resulting in reduced toxicity and significant rise in the rate of germination. This study demonstrated the beneficial use of bacteria of fish gut waste as good biological agents for water treatments, specifically for degradation of textile dye effluents making it suitable to recycle for irrigation purposes.

1 Introduction

Ever since the beginning of humankind, colour has become an important property of expression. Colourants are being used for painting and dyeing of their surroundings, clothes and even skins. Till the middle of the nineteenth century, all colourants were derived from natural sources. With the invention of aniline purple and Tyrian purple with excellent dyeing properties by Perkin in 1896 followed by benzene structure by Kekule strongly stimulated the production of various synthetic dyes leading to replacement of natural dyestuffs with synthetic ones from the beginning of the twentieth century, they were replaced by synthetic dyestuffs (Welham 2000). Worldwide demand for natural dyes is about 10,000 tones. Production of dyestuff and pigments in India is close to 80,000 tons and stands as second largest exporter of dyestuffs and intermediates among developing countries, after China. Textile industry accounts for the largest consumption of dyestuffs, at nearly 80% and the major source of ecotoxic hazard ranging 10–200 mg l^{-1}. The principal route by which dyes enter in the environment is via wastewater. As dyes are designed to be chemically and photolytically stable, they are highly persistent in natural environments.

The International Agency for Research on Cancer (IARC) has classified various dyes like benzidine being associated with cancer (Venturini and Tamaro 1979; Anonym 1982; Mathur et al. 2005). Further, benzidine is known to be carcinogenic to the variety of mammalian species, including humans and reported to induce hepatocellular carcinomas and neoplastic liver nodules in rats after 13 weeks of exposure (Robens et al. 1980). Apart from being phytotoxic (Parshetti et al. 2006), textile dyestuff and wastewaters are detected to be causative for significant reduction in erythrocyte count in the freshwater fish *Gambusia affinis* (Sharma et al. 2009).

Among the commonly used dyes, malachite green (MG) is a multi-organ toxin causing decreased food intake, growth and fertility rates along with damaging liver, spleen, kidney and heart in aquatic species (Werth and Boiteaux 1967; Culp et al. 1999) and irritation to the gastrointestinal tract in humans and permanent injury in eyes (Kumar et al. 2006). Methyl Red (MR) is a commonly used monoazo dye in paper printing and textile dyeing and in the preparation of commercial products (Lachheb et al. 2002). MR is well known to cause eye, skin, respiratory tract and mucous membrane irritation (Mahmoud et al. 2010) and also liver damage upon repeated exposure (Badr et al. 2008). Crystal violet (CV), a triphenylmethane dye, is extensively used in human and veterinary medicine as a biological stain and as a textile dye. It is also used in the preparation of paints and printing. CV is reported to persist in the environment for a long period posing toxic threat to the environment. Therefore, there is an urgent need to develop such eco-friendly and cost-effective biological treatment methods, which can effectively remove the dye from industrial wastewaters for the safety of environment, as well as human and animal health.

1.1 Dye Removal Methods

Textile dyestuff and wastewater are recalcitrant to the degradation. Several physical techniques based on adsorption, osmosis, ultrafiltration, microfiltration and chemical techniques such as coagulation or flocculation combined with flotation and filtration, precipitation/flocculation, electroflotation, electrokinetic coagulation, conventional oxidation irradiation or electrochemical processes were employed to remove colour from the dye containing wastewaters. Several factors, including dye type, wastewater composition, dose or costs of required chemicals, operation costs, environmental fate and handling costs of generated waste products imposed the technical and economic feasibility of each method employed. In general, each technique has its own limitations (Cooper 1993; Maier et al. 2004).

Alternatively, biological treatments are opted to completely degrade dyes (Verma and Madamwar 2003; Moosvi et al. 2005; Pandey et al. 2007; Khalid et al. 2008), ever since microorganisms were found to reduce the dyes by secreting enzymes. Biodegradation methods such as (a) decolourisation using fungi such as *Phanerochaete chrysosporium*, *Trametes* sp. and *Aspergillus* sp, (b) adsorption using live or dead microbial biomass, (c) anaerobic methods to decolourise azo dye using bacteria such as *Bacteroides* sp., *Eubacterium* sp. and *Clostridium* sp. and (d) aerobic oxidatively decolourisation methods. But due to xenobiotic nature, azo dyes are not totally degraded (Laing 1991; Panswald et al. 2001). Moreover, azoreductase and laccase are the two extracellular enzymes that are primarily involved in the bioremediation of textile dyes for which all the microorganisms may not serve as a source.

Fish is highly proteinaceous and easily digestible and a good delicacy. Several fish markets are in operation in each city of Andhra Pradesh. Recent survey revealed that an average minimum 500 kgs of fish waste is generated from each of the fish markets that contain fish scales, fins, gut and other parts of the viscera which is partially used or discarded becoming the main source of pollution and environmental contamination.

Hence in the present study, *Bacillus subtilis* and *Terribacillus saccharophillus* isolated in our laboratory from the gut wastes of freshwater fish, *Labeo rohita*, were examined for their ability to decolourize malachite green. Degraded products were also characterized through, UV-Vis spectroscopy and FTIR and phytotoxic studies.

2   Materials and Methods

2.1   *Materials*

1a. *Labeo rohita*

1b.   Fish-selling shops, Tirupati, Chittoor Dt., A.P

2.   Bacterial Strains
Bacterial strains, *B. subtilis* and *Teribacillus saccharophillus* isolated from the gut of *Labeo rohita*.

B. subtilis   **T. saccharophillus**

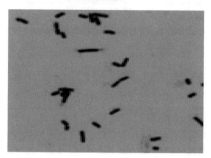
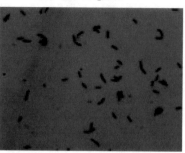

3. The textile dyes such as malachite green (MG) and nutrient broth were obtained from Himedia, India.

Malachite Green

3 Methodology

1. Assessment of Dye Decolourization
   B. subtilis and *T. saccharophillus* were inoculated separately in 250 ml of nutrient broth and incubated at 37 °C for 24 h. After incubation, 25 ml (100 mg/l) of MG was taken, and 5 ml of each bacterial suspension of 10^7 CFU/ml was added separately. The flasks were kept in orbital shaker and incubated at 37 °C. Samples were drawn at 0-, 24- and 48-h intervals centrifuged at 10,000 rpm for 10 min, and the level of decolourization was assessed by measuring absorbance of the supernatant at wavelength maxima (lambda m) of the dye using UV-Vis spectrophotometer.

2. Extraction and Analysis of Dye Degradation Products—UV-Vis spectral analysis
   After complete decolourization of the dye suspended solutions, culture broth was centrifuged at 7000 rpm for 15 min. Equal volume of n-butanol was added to extract metabolites from the supernatant. The extracts were evaporated after removal of aqueous content with anhydrous Na_2SO_4 in a rotary evaporator, dried at 40 °C and used for further analysis.
   The supernatant samples (2 ml; 250 μg/ml) obtained before and after treatment with *B. subtilis* and *T. saccharophillus* were subjected to spectral analysis between

250 and 700 nm. The UV-Vis spectral analysis was carried out using Hitachi UV-Vis spectrophotometer (UV 2900).

Percentage decolourization was calculated from the following formula and tabulated.

$$\% \, \text{Decolorization} = \frac{\text{Initial OD} - \text{Final OD}}{\text{Initial OD}} \times 100$$

3. Fourier Transform Infrared Spectroscopy (FTIR) analysis

   Fourier transform infrared spectroscopy (Bruker ALPHA—T) of samples MG (0 h), 24 and 48 h decolorization products was carried out. The FTIR analysis was done in the mid-IR region of 400–4000 cm^{-1} with 16 scan speed. The samples were mixed with spectroscopically pure KBr in the ratio of 5:95, the pellets were fixed in sample holder, and the analysis was carried out. The transformation of the interferogram into FTIR is carried out mathematically using the software OPUS.

4. Phytotoxicity Studies (Durve et al. 2012)

   Effect of MG on the pattern of seed germination of *Vigna radiate* (Moong) was evaluated. The seeds were germinated in sterile Petri dishes, layered with sterile circular filter paper (diameter 10 cm). Ten seeds were irrigated at room temperature (28 ± 2 °C) for three days with 2 ml of water (control); MG and one of the two bacterial suspensions. Every day 2 ml of respective sample solution was applied to the surface of the filter paper. Each treatment was replicated for three times. Observations were made on daily basis, and rate of germination (%) and root length (cm) were measured.

3.1 Statistical Analysis

Significant differences in the parameters such as shoot and root length were assessed through students' t-test ($p > 0.05$ level).

4 Results and Discussion

1. Dye decolourization

Gradual change in MG colour from bright green to light green up to 48 h after the inoculation of B. *subtilis* and *T. saccharophillus* isolated from the fish gut wastes was observed (Fig. 1).

44/34% decolourization of MG, by B. *subtilis* and *T. saccharophillus* suspended media, respectively, following 24 h of incubation, indicated the ability of these two selected bacteria to rapidly decolourize azo dyes (Fig. 2). Continued exposure up to

(a)   **(b)**

Fig. 1 Malachite green (MG) dye decolourization at 0, 24 and 48 h by **a** *B. subtilis* and **b** *T. saccharophillus*

Fig. 2 Percentage
decolourization of MG with
B. subtilis and *T.
saccharophillus* after 24 and
48 h compared to 0 h

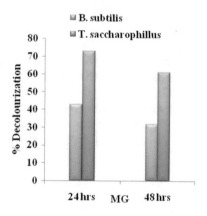

48 h was found to further increase the above-mentioned levels to 73/63%, respectively. In each case compared to *T. saccharophillus* (Fig. 2), *B. subtilis* showed the higher decolourization ability.

Rodriguez et al. as early as in 1999, reported that several industrial dyes were decolourized biocatalytically by extracellular enzymes secreted by microbes. The decolourization effect of different bacterial species may vary. It was reported earlier that the adsorption of chromatophores on the cell surface of the microorganisms results in the decolourization of the dye. The possible mechanism is biosorption which is dependent on functional groups in the dye molecule (Fu and Viraraghavan 2002). Effective decolourization of azo dyes by *Pseudomonas luteola* was reported by Chang and Lin (2000), while Olukanni et al. (2006) reported decolourization by the strain of *Micrococcus*. While Yatome et al. (1991) reported *B. subtilis* to degrade crystal violet, Mabrouk and Yusef (2008) observed its capability in degrading fast red dyes. Malachite green was enzymatically reduced to leucomalachite green and also converted to N-demethylated and N-oxidized metabolites (Cha et al. 2001). Similarly, Pandey and Upadhyay (2010) demonstrated the ability of *Pseudomonas fluorescens* to degrade orange dyes, first by broking it down to 3, 7-diamino-4 hydroxy-naphthalene-2 sulphonic acid sodium salt, which is nontoxic in nature. *Psuedomonas aeruginosa* was observed to cause 78, 76 and 24% decolourization of MR, MG and CV after 24-h incubation by Durve et al. (2012). Recently, Abubacker and Mehala (2014) reported 65, 55 and 50% of decolourization of MG incubated for 96 h with *P. fluorescens*.

2. UV-Visible Spectral Analysis of Dye Degradation

Results of UV-Vis spectroscopic analysis are presented in Fig. 3a, b. Single prominent peak at 615 nm is obtained in UV-Vis spectrum of MG. During absorption or adsorption, it was observed that the intensity of the peak of spectrum decreases with course of time followed by its disappearance with concomitant appearance of some new peaks which were correlated to metabolites produced during degradation. Recording of UV-Vis spectra (200–900 nm) of ethyl acetate extract of MG (250 mg/L) treated with *B. subtilis* showed new peaks at 230 and 330 nm in the present study indicating the degradation of MG resulting in the production of new metabolites in 24 and 48 h (Fig. 3a, b). Further, the appearance of new peaks at 215 and 300 nm in case of treatment with *T. Saccharophillus is* also evincing similar effect. This absorbance was probably a result of impact of similar type of metabolites or/and fragments obtained after degradation of dye molecules by the respective bacteria. These results clearly

Fig. 3 a UV-visible spectra of MG at 0 h (control) and after MG degradation by *B. subtilis* (0, 24; 48 h), *T. saccharophillus* (0, 24 h; 48 h)

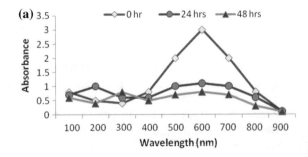

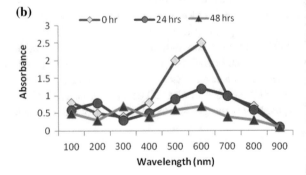

indicate the colour-removing ability of *B. subtilis* and *T. saccharophillus* and hence their potential as biodegrading agents.

In the UV spectra, the decrease in absorbance of the peaks at 220 nm corresponds to benzene ring as demonstrated by Yang and Reilly (1987), and the formation of new peak at 260 nm suggested the reductive destruction of azo-conjugated structure as explained by Feng et al. (2000). The appearance of two new peaks after degradation of MG by *Pseudomonas sps* was shown by Du et al. (2011).

3. Fourier Transform Infrared Spectroscopy (FTIR) analysis

MG (control) and its degraded metabolites are confirmed by FTIR analysis. Remarkable variations were observed in the region 3500–500 cm^{-1} of the FTIR spectroscopy of MG (control), 0-, 24- and 48-h-degraded metabolites of MG by *B. subtilis* and *T. saccharophillus* (Figs. 4a and 5c). FTIR spectra of control MG (control), 0, 24 and 48 h of MG showed the specific peaks in region 2000–500 cm^{-1} representing mono- and para-benzene rings (Tables 1 and 2).

Spectral analysis (FTIR) of MG (0 h) and degraded products (24/48 h) showed functional groups such as alkyl halides, alkenes, carboxylic acids, nitro compounds similar to MG (control). Functional groups such as alcohols and phenols were observed in MG (control), and MG (0 h) was found to be absent after treatment. Extra functional groups such as alpha- and beta-unsaturated aldehydes and ketones were observed in 24-h-treated samples. The variations in the FTIR spectra of MG and those following 24 and 48 h of treatment with *B. subtilis* and *T. saccharophilus* were considered as an indication of biodegradation.

4. Phytotoxicity Studies

Seed germination and plant growth bioassay are the most common techniques used to evaluate the phytotoxicity (Kapanen and Itavara 2001).Water bodies used for irrigation purposes contain untreated effluents from dyeing industry. This practice is of great environmental concern as it is associated with biotic and ecosystem health. Soil fertility is directly and indirectly dependent on irrigation water. Biodegradation of effluent leads to generation of various degradation products. Therefore, it is virtually important to study the toxicity impact of these degradation products on plants (Kalyani et al. 2008).

Phytotoxicity studies using malachite green and its degraded products on *Vigna radiata* are shown in Table 3. The relative sensitivity of *Vigna radiata* towards the MG dyes and its degradation products was studied. 100% germination was observed in the seeds irrigated with tap water (control) (Fig. 6a). Germination of *Vigna radiata* was observed to decrease to 93% with water containing MG dye (Fig. 6b). But germination of *Vigna radiata* seeds soaked in water containing MG treated with *B. subtilis* or with *T. saccharophillus* was 100% (Fig. 6c, d).

The root length in water (control)-treated seeds was 1.5, 4.03 and 9.1 cm on Day 1, 2 and 3, respectively (Table 3), while the same was 0.63, 1.03 and 1.8 cm, respectively, for MG-treated seeds. But following treatment with *B. subtilis* in spite of MG treatment, the seeds showed the root lengths of 1.03, 2.35 and 3.4 cm, respectively. But similar groups treated with *T. saccharophillus* showed root length of 0.85, 2.13

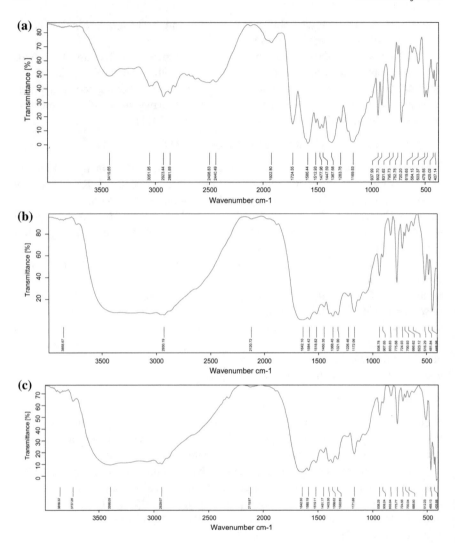

Fig. 4 **a** FTIR spectra of MG (control), **b** FTIR spectra of MG were obtained 0 h by *B. subtilis*, *and* **c** FTIR spectra of MG were obtained 0 h by *T. saccharophillus*

and 2.83 cm respectively (Table 3). Higher root length was observed in seeds grown with treated water compared to the rest of the groups. The results of phytotoxicity studies revealed that the biodegraded products of MG were not toxic to seeds, and hence, the germination process was not affected.

Ren et al. (1996) demonstrated the toxicity of polycyclic aromatic hydrocarbons (PAHs), anthracene (ANT), benzo[a]pyrene (BAP) and fluoranthene to the duck-weed, *Lemna gibba L.* and *Brassica napus L.* seeds. These authors used the germi-nation efficiency, root and shoot growth, and chlorophyll content, as a measurement

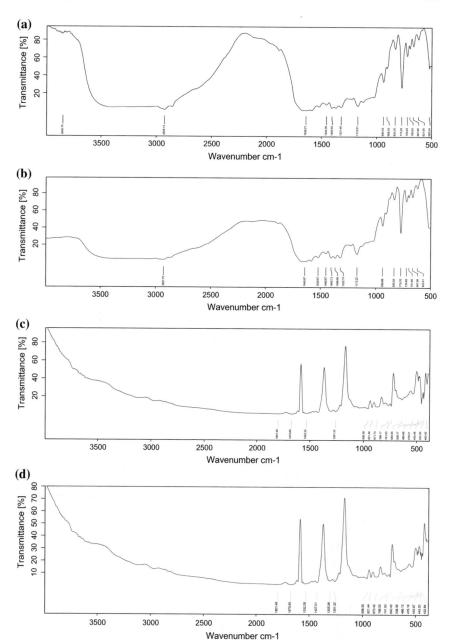

Fig. 5 **a** FTIR spectra of MG were obtained after 24 h by *B. subtilis*, **b** FTIR spectra of MG were obtained after 24 h by *T. saccharophillus*, **c** FTIR spectra of MG were obtained after 48 h by *B. subtilis*, *and* **d** FTIR spectra of MG were obtained after 48 h by *T. saccharophillus*

Table 1 FTIR spectral peaks of malachite green (control) and degraded MG for 0, 24 and 48 h

Samples	Wavelength (cm^{-1})		Vibration	Functional groups
MG (control)	3419.65		O–H stretch, H–bonded	Alcohols, phenols
	3051.95, 2923.44		C–H stretch	Alkanes
	2861.65			
	2498.83			
	2440.49			
	1922.8			
	1724.55		C=O stretch	Alpha-, beta-unsaturated esters
	1586.44		N–H bend	Primary amines
	1521.93, 1477.96		N–O asymmetric stretch	Nitro compounds
	1447.59		C–C stretch (in-ring)	Aromatics
	1367.68		C–H rock	Alkanes
	1283.76		C–H wag	Alkyl halides
	1169.93			
	937.99, 902.70		O–H bend	Carboxylic acids
	832.62, 795.73		C–H "oop"	aromatics
	752.76, 720.20			
	619.65		C–Br stretch	Alkyl halides
	564.15			
	503.37			
MG at 0 h	**B. S**	**T.S**		
	2930.19	2929.07	C–H stretch	Alkanes
		3396.09	O–H stretch, H—bonded	Alcohol, phenols
	2120.72	2115.97	CN stretch	Nitriles
	1642.1	1642.3	– C=C—stretch	Alkenes
	1584.42	1585.19	N–H bend	Primary amines
	1518.82	1519.11	C–C stretch (in-ring)	Aromatics
	1450.35	1451.17 1403.36		
	1368.45	1368.62	C–H rock	Alkanes
	1321.9	1322.89	C–N stretch	Aromatic amines
	1172.06	1171.88	C–C wag	Alkyl halides
	938.78	938.29	O–H bend	Carboxylic acid
	907.65	908.24		

(continued)

Table 1 (continued)

Samples	Wavelength (cm⁻¹)		Vibration	Functional groups
	833.83	833.94		
	775.68	775.77	C–H "oop"	Aromatics
	724.93	724.56		
	700.93	700.04		
	666.62	666.9		
	623.12	512.23	C–Cl stretch	Alkyl halides
	516.29			
MG at 24 h	**B.S**	**T.S**		
	2926.14	2927.7	C–H stretch	Alkanes
	1638.27	1640.87	N–H bend	Primary amines
	1454.39	1520.67	N–O asymmetric stretch	Nitro compounds
	1405.53	1450.67	C–C stretch (ring)	Aromatics
		1405.12		
	1321.43	1322.13	C–N stretch	Aromatic amines
	1172.01	1172.22	C–H wag	Alkyl halides
	940.33	939.86	O–H bend	Carboxylic acids
	908.35	835.63		
	835.2	775.7		
	775.26	724.62	N–H wag	Primary, secondary amines
	725.63	701.85		
	700.93	667.96		
	667.8			
	621.06	622.41	C–Cl stretch	Alkyl halides
	520.24			
MG at 48 h	**B. S**	**T.S**		
	1675.89	1675.65	C=O stretch	Alpha-, beta-unsaturated aldehydes, ketones
	1532.32	1532.28	N–O asymmetric stretch	Nitro compounds
	1261.33	1305.38	C–H wag	Alkyl halides
		1261.32		
	958.38	958.35		
	921.48	921.44		
	871.73	870.42	=C–H bend	Alkenes

(continued)

Table 1 (continued)

Samples	Wavelength (cm^{-1})		Vibration	Functional groups
	768.17	768.05		
	741.63	741.63		
	640.65	640.43	C–Br stretch	Alkylhalides
	538.77	538.36		

Table 2 Functional groups MG before (control) and after degradation treatment with *B. subtilis* and *T. saccharophilus* at 0, 24 and 48 h

MG (control)	MG BS + TS (0 h)	MG BS + TS (24 h)	MG BS + TS (48 h)
Alkyl halides	Alkyl halides	Alkyl halides	Alkyl halides
Alkanes	Alkanes	Alkanes	Alkanes
Carboxylic acid	Carboxylic acid	Carboxylic acid	Carboxylic acid
Aromatic	Aromatic	Aromatic	
Primary amines	Primary amines	Primary amines	
Nitro compounds	Nitro compounds	Nitro compounds	Nitro compounds
	Aromatic amines	Aromatic amines	
Alcohol	Alcohol		
Phenols	Phenols		
Nitriles			
Alpha-, beta-unsaturated esters		Alpha-, beta-unsaturated aldehydes, ketones	

Table 3 Comparison in % germination and root length among the seeds in **a** normal water, **b** water containing malachite green, **c** water containing malachite green treated with *B. subtilis* and **d** water containing malachite green treated with *T. saccharophillus*

Dyes	Germination (%)	Root length (cm)			Average root length (cm)/day
		1st day	2nd day	3rd day	
Water (control)	100	1.5 ± 0.50	4.03 ± 1.4	9.1 ± 2.6	4.96 ± 1.7
MG	93	0.63 ± 0.2	1.03 ± 0.5	1.8 ± 0.7	1.18 ± 0.6
MG + BS	100	1.03 ± 0.61	2.35 ± 0.8	3.4 ± 0.9	2.39 ± 0.9
MG + TS	100	0.85 ± 0.3	2.13 ± 0.8	2.83 ± 1.0	2.19 ± 1.0

Values are mean ± SD of 5 individual observations

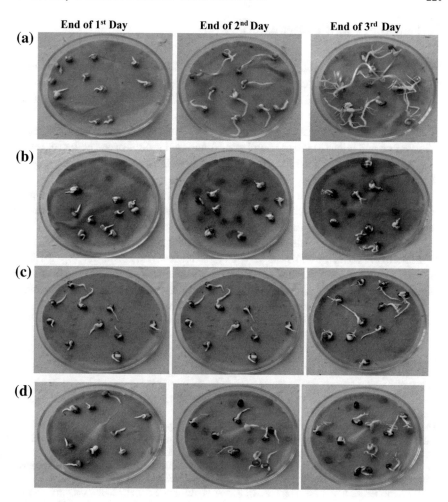

Fig. 6 *Vigna radiata* (Moong) seed germination for 3 days with **a** normal water, **b** water containing malachite green, **c** water containing malachite green treated with *B. subtilis* and **d** water containing malachite green treated with *T. saccharophilus*

for toxicity. Moawad et al. (2003) showed that the textile dyes were toxic to the seed germination process of four plants such as clover, Triticum spp wheat, tomato and lettuce. Though MG is toxic to microbes and plants, its degradation by microorganisms results in significant reduction in its toxicity (Parshetti et al. 2006; Ayed et al. 2009; Chaturvedi et al. 2013). Parshetti et al. (2006) also reported that biodegradation product of MG formed by action of *Kocuria rosea* was nontoxic. Du et al. (2011) also reported that degraded product formed by biodegradation of MG by *Pseudomonas aeruginosa* is nontoxic.

5 Summary and Conclusions

In the present study, *B. subtilis* and *T. saccharophillus* isolated from the gut wastes of freshwater fish; *L. rohita* were found to degrade azo dye.

UV-Vis studies showed that *B. subtilis* and *T. saccharophillus* have a significant ability to degrade azo dye such as malachite green.

Fourier transform infrared spectroscopy (FTIR) studies demonstrated that remarkable change and reduction in the number of functional groups occur with progress in treatment time clearly illustrating the impact of both the bacteria on the degradation of malachite green.

Phytotoxicity studies further clearly showed that addition of *B. subtilis* and *T. saccharophillus* to the growth medium of seeds can successfully increase the rate of germination through degrading integrity of the dye.

Thus, this study clearly demonstrated that the fish gut bacterial isolates such as *B. subtilis* and *T. saccharophillus* can serve as good tools to treat wastewaters, especially the textile dye effluents. Further, these wastes which otherwise are thrown into water bodies or dumped in open lands leading to additional pollution can well be avoided and explored for their reutilization using probiotic microbes.

References

Abubacker, M. N., & Mehala, T. (2014). Physico-chemical analysis of textile dye effluent using microbial consortia mediated degradation process.

Anonym—International Agency for Research on Cancer, Suppl. 4 (1982): IARC Monographs on the evaluation of the carcinogenic risk of chemicals to humans, chemicals, industrial processes and industries associated with cancer in humans. Lyon: IARC.

Ayed, L., Chaieb, K., & Cheref, A. (2009). Biodegradation of triphenylmethane dye malachite green by Spingomonas paucimobilis. *World Journal Microbial Biotechnology, 25,* 705–711.

Badr, A., Müller, K. J., Schäfer-Pregl, R., El Rabey, H., Effgen, S., et al. (2008). On the origin and domestication history of barley (*Hordeum vulgare*). *Molecular Biology Evolution, 17,* 499–510.

Cha, C. J., Doerge, D. R., & Cerniglia, C. E. (2001). Biotransformation of Malachite Green by the FungusCunninghamella elegans. *Applied and Environmental Microbiology,* 4358–4360.

Chang, J-S., Lin, Y-C. (2000). Fed-batch bioreactor strategies for microbial decolorization of azo dye using a *Pseudomonasluteola* strain. *Biotechnol Progress, 16,* 979–985.

Chaturvedi, A. K., Anderson, W. F., Lortet-Tieulent, J., et al. (2013). Worldwide trends in incidence rates for oral cavity and oropharyngeal cancers. *Journal of Clinical Oncology, 31,* 4550–4559.

Cooper, P. (1993). Removing color from dyehouse wastewaters a critical review of technology available. *Journal of the Society of Dyers and Colourists, 109,* 97–101.

Culp, S. J., Blankenship, L. R., Kusewitt, D. F., Doerge, D. R., Mulligan, L. T., & Beland, F. A. (1999). Toxicity and metabolism of malachite green and leucomalachite green during short-term feeding to Fischer 344 rats and B6C3F1 mice. *Chemico Biological Intreactions, 1223*(153), 170.

Du, L. N., Wang, S., Li, G., Wang, B., Jia, X. M., Zhao, Y. H., & Chen, Y. L. (2011). Biodegradation of malachite green by Pseudomonas sp. strain DY1 under aerobic condition: characteristics, degradation products, enzyme analysis and phytotoxicity. *Ecotoxicology, 20*(2), 438–446.

Durve, A. A., Gupta, A. R., & Naphade, S. R. (2012). Decolourisation of textile dyes and biological stains by bacterial strains isolated from Industrial effluents. *Advances in Applied Science Research, 3*(5), 2660–2671.

Durve, A., Naphade, S., Bhot, M., Varghese, J., & Chandra, N. (2012). Characterisation of metal and xenobiotic resistance in bacteria isolated from textile effluent. *Advances in Applied Science Research, 3*(5), 2801–2806.

Feng, W., Nansheng, D., & Helin, F. L. (2000). Degradation mechanism of azo dye Cl reactive red 2 by iron powder reduction and photooxidation in aqueous solutions. *Chemosphere, 41*, 1233–1238.

Fu, Y., & Viraraghavan, T. (2002). Dye biosorption sites in *Aspergillus niger*. *Bioresource Technology, 82*, 139–145.

Kalyani, D. C. P. S., Patil, J. P., & Jadhav, S. P. Govindwar. (2008). Biodegradation of reactive textile dye Red BLI by an isolated bacterium Pseudomonas sp. SUK1. *Bioresource Technology, 99*, 4635–4641.

Kapanen, A., & Itavaara, M. (2001). Ecotoxicity tests for compost applications. *Ecotoxicology and Environmental Safety, 49*, 1–16.

Khalid, A., Arshad, M., & Crowley, D. E. (2008). Decolorization of azo dyes by Shewanella sp. Under saline conditions. *Applied Microbiology and Biotechnology, 79*, 1053–1059.

Kumar, K. V., Ramamurthi, V., & Sivanesan, S. (2006). Biosorption of malachite green, a cationic dye onto Pithophora sp., a fresh water algae. *Dyes and Pigments, 69*, 102–107.

Lachheb, H., Puzenat, E., Houas, A., Ksibi, M., & Elaloui, E. (2002). Photocatalytic degradation of various types of dyes (alizarin S, crucein orange G, methyl red, Congo red, methylene blue) in water by UV-irradiated Titania. *Catalysis B Environmental, 39*, 75–90.

Laing, I. G. (1991). The impact of effluent regulations on the dyeing industry. *Review of Progress in Coloration and Related Topics, 21*(1), 56–71.

Mabrouk , M. E. M., & Yusef, H. H. (2008). Decolourization of fast red by Bacillus subtilis. *Journal of Applied Sciences Research, 4*(3), 262–269.

Mahmoud, M. A., Kastner, A., & Yeboah, J. (2010). Antecedents, environmental moderators and consequences of market orientation: A study of pharmaceutical firms in Ghana. *Journal of Medical Marketing, 10*(3), 231–244.

Maier, J., Kandelbauer, A., Erlacher, A., Cavaco-Paulo, A., & Gubitz, M. G. (2004). A new alkali thermostable azoreductase from Bacillus sp. Strain SF. *Applied and Environmental Microbiology, 702*, 837–844.

Mathur, N., Hatnagar, P. B., & Bakre, P. (2005). Assessing mutagenicity of textile dyes from pali (Rajasthan) using Ames bioassay. *Applied Ecology and Environmental Research, 4*(1), 111–118 further.

Moawad, H., Abd el-Rahim, W. M., & Khalafallah, M. (2003). *Journal Basic Microbiology, 43*(3), 218–229.

Moosvi, S., Keharia, H., & Madamawar, D. (2005). Decolorization of textile dye reactive Violet 5 by a newly isolated bacterial consortium RVM11. *World Journal of Microbial Biotechnology, 21*, 667–672.

Nirmalarani, J., & Janardhanan, K. (1988). Effect of South India Viscose factory effluent on seed germination seedling growth and chloroplast pigments content in five varieties of Maize (Zea mays I). *Madras Agricultural Journal, 75*, 41.

Olukanni, O. D., Osuntoki, A. A., & Gbenle, G. O. (2006). Textile effluent biodegradation potentials of textile effluent adapted and non-adapted bacteria. *African Journal of Biotechnology, 5*, 1980–1984.

Pandey, B. V., & Upadhyay, R. S. (2010). Pseudomonas fluorescens can be used for bioremediation of textile effluent Direct Orange-102. *Tropical Ecology, 51*(2S), 397–403.

Pandey, A., Singh, P., & Iyengar, L. (2007). Bacterial decolorization and degradation of azo dyes. *International Biodeterioration and Biodegradation, 59*, 73–84.

Panswald, T., Lamsamer, K., & Anotai, J. (2001). Decolorization of azo-reactive dye by polyphosphate and glycogen-accumulating organisms in an anaerobic-aerobic sequencing batch reactor. *Journal of Bioresource Technology, 76*, 151–159.

Parshetti, G., Kalme, S., Saratale, G., & Govindwar, S. (2006). Biodegradation of Malachite green by *Kocuria rosea* MTCC 1532. *Acta Chimica Slovenica, 53*, 492–498.

Perkin, W. H. (1896). *Journal of Chemical Society*, 596.

Ren, L., Zeiler Lorelei, F., George Dixon, D., & Bruce, M. (1996). Photoinduced effects of polycyclic aromatic hydrocarbons on *Brassica napus* (Canola) during germination and early seedling development. *Greenberg1 Ecotoxicology and Environmental Safety, 33*, 73–80.

Robens, J. F., Dill, G. S., Ward, J. M., Joiner, J. R., Griesemer, R. A., & Douglas, J. F. (1980). Thirteen-week subchronic toxicity studies of Direct Blue 6, Direct Black 38, and Direct Brown 95 dyes. *Toxicology Applied Pharmaceutical, 54*, 431–442.

Rodriguez, C. R. (1999). A *Saccharomyces cerevisiae* RNA 5'-triphosphatase related to mRNA capping enzyme. *Nucleic Acids Research, 27*(10), 2181–2188.

Sharma, S., Sharma, S., Singh, P. K., & Swami, R. C. (2009). Exploring fish bioassay of textile dye wastewaters and their selected constituents in terms of mortality and erythrocyte disorders. *Bulletin of Environmental Contamination and Toxicology, 83*, 29–34.

Venturini, S., & Tamaro, M. (1979). Mutagenicity of anthraquinone and azo dyes in Ames' *Salmonella typhimurium* test. *Mutation Research, 68*, 307–312.

Verma, P., & Madamwar, D. (2003). Decolorization of synthetic dyes by a newly isolated strain of *Serratia maerascens. World Journal Microbiology Biotechnology, 19*, 615–618.

Welham, A. (2000). The theory of dyeing (and the secret of life). *Journal of the Society of Dyers and Colourists, 116*, 140–143.

Werth, G., & Boiteaux, A. (1967). The toxicity of the triphenylmethane dyestuff malachite green, as an uncoupler of oxidative phosphorylation in vivo and in vitro. *Arch fur Toxical, 23*, 82–103.

Yang, M., & Reilly, J. P., (1987). A reflectron mass spectrometer with UV laser induced surface ionization. *International journal of mass spectrometry and ion processes, 75*, 209–219.

Yatome, C., Ogawa, T., & Matsui, M. (1991). Degradation of Crystal Violet by *Bacillus subrilis. Journal of Environment Science Health, A26*, 75–87.

Utilization of Rice Husk Ash for Defluoridation of Water

C. M. Vivek Vardhan and M. Srimurali

Abstract Effective utilization and safe disposal of solid wastes is one of the major challenges placed before the modern-day scientist, in the wake of exploding population. Rice husk is a major agricultural waste being generated by an agricultural nation such as India, and therefore warrants proper utilization before disposal. On the other hand, fluorosis is a burning problem caused due to consumption of groundwater consisting of excessive fluoride, prevalent in several parts of the world. Therefore, in this study, Rice Husk Ash was impregnated with Lanthanum and investigated for defluoridation of water by adsorption. Batch adsorption experiments were conducted for fluoride removal using Lanthanum-Impregnated Rice Husk Ash (LIRHA) and tested for various parameters such as pH, time of agitation, influence of ions, and optimum dosage of sorbent. It was observed that LIRHA removed fluoride to less than the permissible limit in the naturally encountered pH of water. The optimum time and dosage of LIRHA were found to be 240 min and 6 g/L, respectively. The anions phosphates and chlorides were found to be detrimental for fluoride sorption probably due to competitive of those ions with fluoride on the active sorption sites on LIRHA. Thus, this study revealed that the agricultural waste, Rice husk can be effectively utilized for removal of fluoride from water and subsequently disposed off safely.

Keywords Adsorption · Defluoridation · Water · Low-cost · Lanthanum

C. M. Vivek Vardhan (✉)
Department of Civil Engineering, Malla Reddy Engineering College (A), Maisammaguda, Medchal, Hyderabad 500100, Telangana State, India

M. Srimurali
Department of Civil Engineering, Sri Venkateswara University College of Engineering, S.V. University, Tirupati, India

© Springer Nature Singapore Pte Ltd. 2020
S. K. Ghosh et al. (eds.), *Recent Trends in Waste Water Treatment and Water Resource Management*, https://doi.org/10.1007/978-981-15-0706-9_21

1 Introduction

India being an agricultural nation generates huge quantities of Rice husk as agricultural waste. About 24 million tons of Rice husk is being generated by the nation annually (Pandey et al. 2012). A minor portion of this waste is used as cattle feed and in manufacture of cement and refractory bricks. The remaining major portion is dumped, which results in its decomposition and subsequent emission of methane gas. Therefore, an effective way for utilizing rice husk has to be found and subsequently disposed safely. Concomitant to this challenge, contamination of groundwater with fluoride and the resulting fluorosis is a major threat faced by mankind. Fluoride has dual impacts on living organisms. In lower concentrations, fluoride is very beneficial as it assists in formation of bones and prevents dental decay, whereas in higher concentrations, it causes serious health issues such as skeletal fluorosis, dental fluorosis, mental derangements, dwarfishness, etc. Therefore, it is imperative that fluoride has to be removed from water before consumption. The World Health Organization (WHO) has stipulated a maximum permissible limit of 1.5 mg/L of fluoride in drinking water. Various technologies are available for defluoridation of water and are partially successful to various extents. But, as fluorosis is not yet completely eradicated, the focus has recently shifted to use of rare earth materials as adsorbents for defluoridation with a fairly reasonable degree of success (Vardhan and Srimurali 2018; Vivek Vardhan and Srimurali 2015, 2016). However, research is still deficient on achieving a dual advantage of simultaneously utilizing waste and removing fluoride from water. Therefore, in this study, an attempt was made to employ Rice Husk Ash impregnated with Lanthanum for removal of fluoride from water. The main objectives of this study are to: (1) Impregnate Lanthanum onto Rice Husk Ash, (2) Find the optimum time of agitation and the dosage of sorbent required, (3) Investigate the influence of pH on fluoride removal, and (4) Study the impact of anions on defluoridation.

2 Materials and Methods

2.1 Chemicals

All reagents used in the present investigation were procured from E. Merck Ltd., India and were of analytical reagent grade. Sodium fluoride of 221 mg was dissolved in distilled water and made up to 1 L in a volumetric flask to prepare a stock fluoride solution of 100 mg/L. Stock solution was diluted appropriately to obtain fluoride solutions of required working concentrations.

2.2 Adsorbent

Rice Husk Ash (RHA) was procured from a local rice mill and was sieved to a mean size of 475 μm.

2.3 Adsorbent Preparation

Lanthanum carbonate of varying weights was mixed with 0.05 L of distilled water and 0.1 N HCl was added to it drop wise until Lanthanum carbonate got dissolved completely. To this solution, 0.1 N NaOH was added under constant stirring until precipitates were visually observed. RHA of mean size 475 μm and 20 ± 0.2 g weight was added to this mixture. This mixture was mixed for 6 h in a magnetic stirrer. Subsequently, it was decanted, dried, and heated at 300 °C for 4 h in a muffle furnace. It was thoroughly washed with distilled water and dried. The obtained material is Lanthanum-Impregnated Rice Husk Ash (LIRHA).

2.4 Batch Adsorption Experiments

Teflon flasks of 250 mL capacity and a working volume of 100 mL were used throughout the experiments. Laboratory spiked fluoride solutions of 10 ± 0.2 mg/L were prepared and adsorbent of required dosage was added to each flask. These flasks were agitated in a rotary shaker of make Kaizen imperial. The optimum rate of agitation for highest removal of fluoride was determined by varying the speed of agitations at 40, 80, 120, 160, and 200 rpm, at 25 ± 1 °C. Highest fluoride removal was observed at 160 rpm and so all experiments were conducted at this speed of agitation. After agitating for stipulated time periods, the contents were withdrawn, filtered in a Whatman filter paper of size 42, and analyzed for fluoride removal using SPADNS method (APHA 2005). Fluoride was analyzed using a spectrophotometer, Evolution 201, of Thermo Scientifics. pH adjustments were carried out using 0.1 N NaOH and 0.1 N H_2SO_4. Individual ions of Cl^-, SO_4^{2-}, PO_4^{3-}, HCO^{3-}, and NO^{3-} were added in concentration of 250 ± 5 mg/L in addition to fluoride ions to evaluate its impact. Concentration of Lanthanum was measured using an atomic adsorption spectrophotometer AAS, GBS 932 plus.

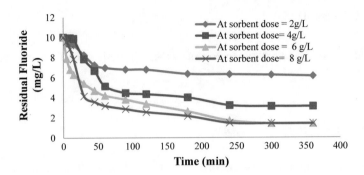

Fig. 1 Influence of time of agitation and dose of LIRHA on defluoridation

3 Results and Discussion

3.1 Time and Dose Studies of RHA and LIRHA

Experiments were carried out to arrive at the optimum time and dose required for adsorption to take place using LIRHA and the results are presented in Fig. 1. It can be observed that the time and dose of sorbent required for defluoridation using LIRHA are 240 min and 6 g/L, respectively, which left a residual fluoride of 1.39 mg/L.

3.2 Influence of pH

Role of pH on defluoridation using LIRHA was examined and the results are presented in Fig. 2. It can be observed that LIRHA performed better in removing fluoride from water in a pH range of 7–9. Formation of HF ions at lower pH values and OH

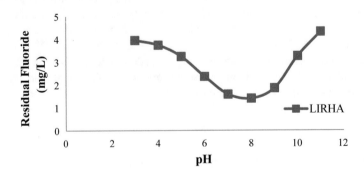

Fig. 2 Influence of pH of solution on fluoride removal by LIRHA

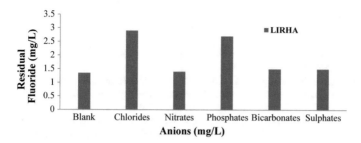

Fig. 3 Influence of anions on sorption of fluoride by LIRHA

ions at higher pH values could be the reason for lesser defluoridation below and above the optimum pH ranges observed.

3.3 Influence of Anions

Efficiency of fluoride removal by LIRHA was tested in the presence of the anions Cl^-, SO_4^{2-}, PO_4^{3-}, HCO^{3-}, and NO^{3-}, each individually and the results are presented in Fig. 3. It can be observed that chlorides and phosphates impeded sorption of fluoride by LIRHA, whereas the ions nitrates, bicarbonates, and sulfates exhibited a lesser influence. Similar trend of competition of phosphates with fluorides for active sorption sites was reported by Bia et al. (2012) on ion modified montmirillonite. Chlorides were also reported to be highly competitive against fluoride for sorption on expanded clay aggregate, according to Sepehr et al. (2014).

4 Concluding Remarks

The aim of this work is to simultaneously utilize the agricultural waste Rice Husk Ash and to remove fluoride from water.

Agitated non-flow batch studies demonstrated that LIRHA is capable of removing fluoride from water to less than the permissible limits.

Normally encountered pH range was found to be favorable for fluoride removal by LIRHA.

It was observed that the time and dose of LIRHA required for defluoridation were 240 min and 6 g/L, respectively.

Presence of anions such as phosphates and chlorides exerted a detrimental effect on removal of fluoride from water using LIRHA, possibly due to competition of ions.

These investigations revealed that the agricultural waste Rice Husk can be utilized toward removal of fluoride from water and should be disposed off safely.

References

APHA. (2005). *Standard methods for the examination of water and wastewater* (19th ed.). Washington, DC: APHA, AWWA, WEF.

Bia, G., De Pauli, C. P., & Borgnino, L. (2012). The role of Fe(III) modified montmorillonite on fluoride mobility: adsorption experiments and competition with phosphate. *Journal of Environmental Management, 100,* 1–9. https://doi.org/10.1016/j.jenvman.2012.01.019.

Pandey, R. S., Sar, K., & Bhui, A. K. (2012). Feasibility of installating rice husk power plant in Chhattisgarh to meet sustainable energy demands. *International Journal of Advanced Engineering Research and Studies* (2017 May).

Sepehr, M. N., et al. (2014). Defluoridation of water via light weight expanded clay aggregate (LECA): Adsorbent characterization, competing ions, chemical regeneration, equilibrium and kinetic modeling. *Journal of the Taiwan Institute of Chemical Engineers, 45*(4), 1821–1834. https://doi.org/10.1016/j.jtice.2014.02.009.

Vardhan, V., & Srimurali, M. (2018). Preparation of lanthanum impregnated pumice for defluoridation of water: Batch and column experiments. *Journal of Environmental Chemical Engineering.*

Vivek Vardhan, C. M., & Srimurali, M. (2015, January). Defluoridation of drinking water using a novel sorbent: Lanthanum-impregnated green sand. *Desalination and Water Treatment, 3994,* 1–11. http://www.tandfonline.com/doi/abs/10.1080/19443994.2015.1012330.

Vivek Vardhan, C. M., & Srimurali, M. (2016). Removal of fluoride from water using a novel sorbent lanthanum-impregnated bauxite. *SpringerPlus, 5*(1), 1426. http://springerplus.springeropen.com/articles/10.1186/s40064-016-3112-6.

Comparison of the Suitability of Activated Carbon Derived from *Mangifera indica* by Four Physicochemical Methods for the Removal of Methylene Blue Dye

Moumita Biswas, Anirban Ghosh and Saswata Bose

Abstract Activated carbon (AC) from mango (*Mangifera indica*) peels was developed through four environmentally benign methodologies. Activated carbons were prepared by both thermal and chemical activation and termed mango peel activated carbon (P-MPAC). One involving a single physical activation was termed P-MPAC. Two methods, involving the two-step activation process where chemical activation (using H_2SO_4) was followed by physical activation. One of the two variants activated at 300 °C was termed as C-300-MPAC and the other activated at 700 °C was termed C-700-MPAC. One has an additional physical activation before the two-step activation process was named PCP-MPAC (physical 300 °C—chemical–physical 700 °C three activation steps). Thermal activation was done at two temperatures 300 and 700 °C to explore the effect of high- and low-temperature thermal activation in generating pores on the surface of the AC. The four variants (namely P-MPAC, C-300-MPAC, C-700-MPAC, and PCP-MPAC) were explored to determine the ideal method of AC preparation. Their ability as an adsorbent for the removal of a basic dye methylene blue was examined. The pH and adsorbate concentration was optimized to get ideal conditions for the removal of the two respective dyes from their respective aqueous solutions. ACs activated at 700 °C performed significantly better at removing dyes from their aqueous solution. Three-step activation showed an enhanced effect in removing the impurity.

Keywords International society of waste management · Air and water

1 Introduction

Mango (*Mangifera indica* L., family Anacardiaceae) is a fruit native to South Asia. It is most popular for its consumption for its juicy pulp which regards it the name "King of fruits" in India. India is also the largest producer of the juicy fruit producing an

M. Biswas · S. Bose (✉)
Chemical Engineering Department, Heritage Institute of Technology, Kolkata, India
e-mail: saswata.bose@jadavpuruniversity.in

A. Ghosh · S. Bose
Chemical Engineering Department, Jadavpur University, Kolkata, India

© Springer Nature Singapore Pte Ltd. 2020
S. K. Ghosh et al. (eds.), *Recent Trends in Waste Water Treatment and Water Resource Management*, https://doi.org/10.1007/978-981-15-0706-9_22

outstanding 40% of the total produce worldwide which stands at an estimated amount of 46 million metric tons as of 2016. India thus produces a high amount of organic waste rich in functional group in the form of mango peels and seeds which have little use as a delicacy but can be tapped into as a very good source of producing resource from waste (FAOSTAT 2005).

The edible part (the pulp), which makes up to 33–85% of the fresh fruit depending on the various breeds of the fruit finds use both in households and commercially. Household use is mostly seasonal in nature, i.e. during the fresh fruit comes into the market, but commercial uses run throughout the year. Household uses mainly include consuming the fruit raw due to its one of a kind flavour and also making pickles from the dried fruit. The major commercial products made from the fruit are milkshakes, drinks, and juices. Brand names such as Maaza and Slice which specializes in mango-related beverages are hardly unknown to any in India. Processed products such as nectars, concentrates, jam, jelly powders, fruit bars, and dried mango products also require a huge amount of the fruit produce (Wu et al. 1993).

Processing of the fruit entails that 7–24% of the entire weight of the fruit which constitutes the mango peels turns into waste. Mango peels mostly end up turning into a stockpile of waste that remains unused that serves no end. Although possessing chemical characteristics, which indicate a rich source of bioactive compounds such and enzymes such as protease, peroxidase, polyphenol oxidase, and carotenoids use of these essential organic structures have not yet been found. There is also an abundant content of vitamins C and E, dietary fibres, enzymes, and carbohydrate content of 20.80–28.20% in dry weight samples of mango peel (Ajila et al. 2007). Despite these beneficial traits not only has there been any suitable application of them but also these rich organic produce end up being incinerated or being left to putrefy. Land usage is hampered and the over hygiene of the environment in which such commercial usage of mango is being carried out is jeopardized. Presence of functional groups such as carboxylic (pectin) and hydroxyl groups (cellulose) is reported in the polymers contained in abundance in the mango peel matrix. The presence of these functional groups increases the chance of mango peels being favoured as suitable material for the production of AC (Pehlivan et al. 2008; Shukla and Pai 2005; Xuan et al. 2006).

In the current condition of the planet, where all the resources including the land, water bodies, and air are getting polluted beyond imagining, saving the environment is the need of the hour. Many water purifying technologies are available to mankind for use such as electro-coagulation, membrane filtration, electrochemical destruction, ion exchange, irradiation, advanced oxidation, ozonation, precipitation but none prove to be as cheap and effective as the method of adsorption (López-Grimau and Gutiérrez 2006). Carbonaceous materials such as AC prove to be some of the most effective adsorbents having relatively lower production costs too. An application of the as-obtained industrial by-product(waste)-mango peels would be saving a lot of resources. Currently, activated carbon is mostly derived from charcoal, wood derivatives bones, and nutshells. The option and scope of producing ACs from organic sources are wide and endless. Attempts are being made worldwide to produce AC from organic wastes to reduce the load on the environment (Lakshmi et al. 2009; Lillo-Rodenas et al. 2003). An application like producing AC would serve the dual purpose of not only help reducing the load on the land used by commercial users of mango pulp but also contribute to the cleaning up of water by being used as an adsorbent to remove the pollutants to produce safe water for drinking and recreational processes.

The main aim of work is to design a suitable adsorbents derived from mango peels, to test their effectiveness in the removal of dyes, and to explore treatment methods during production which give rise to desired traits in an AC towards removal of dyes. Many organic sources have been tried as sources for the production of AC but mango peels irrespective of its chemical properties are not a much explored precursor for AC production. Methylene blue dyes were removed using the ACs prepared by the four respective routes of mango peel activated carbon (MPAC) production.

2 Materials and Methods

2.1 Preparation of AC

Local markets of Kolkata were selected as the source of the MPs. The as-obtained MPs contained juice of mangoes and were not purified. Thus, a two-step washing procedure was adopted. This consisted of thorough and repeated washing by tap water after which washing by double distilled water ensued. This step was followed by two drying steps; first of all, the sun drying for one day followed by oven drying at 100 °C for 24 h. Drying was stopped at the stage when the mango peels obtained a nice crunchy texture ideal for hand grinding. This is followed by thorough grinding of the mango peels which resulted in fine particles of powdery material named as mango peel grinded powder (MPGP).

2.1.1 Thermal Activation

MPCP the powder was taken in several silica crucibles in measured amounts and carbonized ensued at 700 °C for 2 h in a muffle furnace. The product obtained from the muffle furnace after thermal activation was sieved through a 250 μm mesh to obtain the AC. The AC so produced was called as temperature activated-mango peel activated carbon (P-MPAC).

2.1.2 Chemical Activation

First Process

MPCP was taken in glass petri dish and mixed thoroughly till with concentrated H_2SO_4 in 1:1 weight ratio till uniformity was obtained. This mixture was kept overnight (24 h) under stirring at room temperature. The product thus obtained was then washed with double distilled water repeatedly till neutral pH was reached. The powder after neutralization was then transferred into silica crucibles and carbonized at 700 °C for 2 h in a muffle furnace. The product obtained from the muffle furnace after thermal activation was sieved through a 250 μm mesh to obtain the AC. C-700-MPAC was the name given to the as-produced AC.

Second Process

MPCP was taken in petri dish and mixed thoroughly with concentrated H_2SO_4 in 1:1 weight ratio till uniformity was obtained. This mixture was kept overnight (24 h) under stirring at room temperature. The product thus obtained was then washed with double distilled water repeatedly till neutral pH was reached. The powder after neutralization was then transferred into silica crucibles and carbonized at 300 °C for 3 h (as we maintain temperature well below 700 °C, we have raised the time of carbonization to 3 h) in a muffle furnace. The product obtained from the muffle furnace after thermal activation was sieved through a 250 μm mesh to obtain the AC. C-300-MPAC was the name given to the as-produced AC.

Third Process

MPCP was taken in a silica crucible and carbonized at 300 °C for 2 h. The obtained product was crushed thoroughly and mixed with equal proportion of concentrated H_2SO_4 in 1:1 weight ratio. This mixture was kept overnight (24 h) under stirring at room temperature. The product thus obtained was then washed with double distilled water repeatedly till neutral pH was reached. The powder after neutralization was then transferred into silica crucibles and carbonized at 700 °C for 1 h in a muffle furnace. The product obtained from the muffle furnace after thermal activation was sieved through a 250 μm mesh to obtain the AC. PCP-MPAC was the name given to the as-produced AC.

2.2 *Preparation of Adsorbate*

The 10 μM solution of methylene blue dye solution was prepared to be used as the adsorbate in the adsorption studies. 1.6 mg of methylene blue of analytical grade was added to 500 ml of double distilled water to prepare the 10 μM solution of methylene blue.

2.3 Reagents and Analytical Measurements

The reagents used were of analytical grade, they were supplied from EMPARTA. The adsorbate used was methylene blue was supplied by SRL (Shanghai Richem International). H_2SO_4 used was of 98% purity. Mango peels were obtained from local markets in Kolkata. Double distilled water was used in all solutions.

3 Effect of pH

One of the most fundamental parameters that effect how efficiently an adsorption process will perform is the pH of the solution. The charges present on the surface of the adsorbent and the electrical interactions between the adsorbent and adsorbate is greatly affected by the pH of the solution. Speciation and degree of ionization are also affected by the pH of a solution. The effect of pH of solution on adsorption of dyes (concentration $= 10 \mu M$) using MPAC ($m = 1$ g/l) was evaluated at different pH varying from 1 to 11 at 303 K for six hour and the results are plotted in Fig. 1. It could easily be understood from the figure that low pH of 3 is the most favourable pH for carrying out the further adsorption studies as the highest percentage of removal was obtained at it.

Point of zero charge (PZC) is defined as the pH for which the adsorbent surface bears no charge. Any increase of pH from the PZC will render the surface negatively charge and any decrease from the PZC will make the surface positively charged. Thus, PZC is the optimum for the purpose of adsorption studies. It is clear from the figure MPAC must have a PZC of 3 for methylene blue dye and act neutral at this pH.

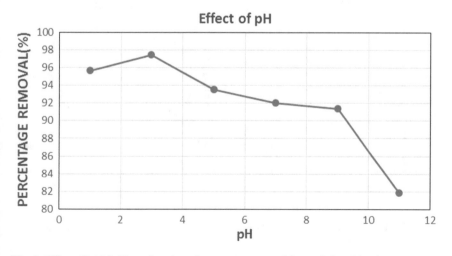

Fig. 1 Effect of initial pH as a function of percentage removal for methylene blue dye

3.1 Batch Kinetic Study

Experimental study was conducted in batch mode for exploring the ability of the MPACs for the adsorption of methylene blue dye from its aqueous solution. The experiments were performed in Erlenmeyer flasks containing 50 nl of dye solution. 50 mg of the prepared adsorbents were added to the prepared 10 micromolar solution of methylene blue dye. Stirring was conducted at 500 RPM. Initial pH of the solution was changed using 0.1 N NaOH/0.1 N HCl. Filtrate analysis was carried out using UV-VIS spectrophotometer at $\lambda_{max} = 554$ nm for methylene blue. The dye adsorbed per unit weight of MPACs and percentage removal of Rhodamine B was calculated using following equations (Fig. 2):

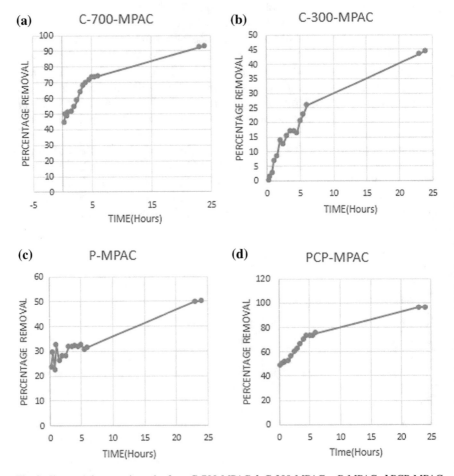

Fig. 2 % removal versus time plot for **a** C-700-MPAC; **b** C-300-MPAC; **c** P-MPAC; **d** PCP-MPAC

$$\text{Adsorption Capacity}\,(q_e) = \frac{(C_0 - C_e)V}{m}$$

$$\%\,\text{Removal} = \frac{(C_0 - C_e) * 100}{C_0}$$

where

C_e Residual or equilibrium concentration of dye solution (mg/l)
C_o Initial concentration of dye solution (mg/l)
m Amount of MPAC (g)
q_e Amount of dye adsorbed on BPAC/adsorption capacity (mg/g)
V Volume of dye solution (l).

4 Adsorption Kinetics Results

The results of the kinetic study along with the predicted model (shown in bold text in Table 1) are listed in Figs. 3 and 4.

5 Discussion

The inference that can be drawn from the batch study is that PCP-MPAC (3.221 mg/g) and C-700-MPAC (3.0931 mg/g) having the mentioned adsorption capacities are much more superior as adsorbents than C-300-MPAC and P-MPAC having adsorption capacities 1.4291 mg/g and 1.6815 mg/g, respectively.

The enhanced ability of C-700-MPAC as adsorbent compared to the P-MPAC despite having a similar thermal treatment might be attributed to the fact that chemical activation was performed on the C-700-MPAC which can lead to the generation of oxygenated groups on the surface of the ACs. The oxygenated groups may play a crucial role in the adsorption process through chemisorption. The dual effect of high-temperature physical activation and chemisorption could lead to an enhancement of the adsorption capacity of C-700-MPAC.

P-MPAC although received a similar treatment as C-700-MPAC thermally the lack of the chemical treatment could be held responsible for the low adsorption capacity. Chemical treatment by a strongly oxidizing chemical such as concentrated sulphuric acid can generate additional oxidizing functional groups which aid in the adsorption of the dye from its aqueous solution. It can also be seen in the case of C-300-MPAC that although it received both chemical and physical treatment it is the temperature of activation which made a huge difference in adsorption capacities. It may be inferred that the low temperature of 300 °C is not sufficient to complete the thermal cracking of the precursor mango peels.

Table 1 Batch data of the adsorbents

Name of sample	Type of kinetics	R^2	Error (%)	K	Model predicted Q_e (mg/g)
PCP-MPAC	Pseudo-first-order kinetics	0.9785	54.38	$0.2616\ h^{-1}$	1.4063
	Pseudo-second-order kinetics	**0.9943**	**4.48**	**0.2241 g/mg h**	**3.221**
P-MPAC	Pseudo-first-order kinetics	0.9245	34.31	$0.2024\ h^{-1}$	1.0531
	Pseudo-second-order kinetics	**0.976**	**4.8**	**0.3108 g/mg h**	**1.6815**
C-300-MPAC	Pseudo-first-order kinetics	0.9917	15.53	$0.1593\ h^{-1}$	1.2071
	Pseudo-second-order kinetics	**1**	**0**		**1.4291**
C-700-MPAC	Pseudo-first-order kinetics	0.9956	57.62	$0.2010\ h^{-1}$	1.2606
	Pseudo-second-order kinetics	**0.996**	**3.98**	**0.2588 g/mg h**	**3.0931**

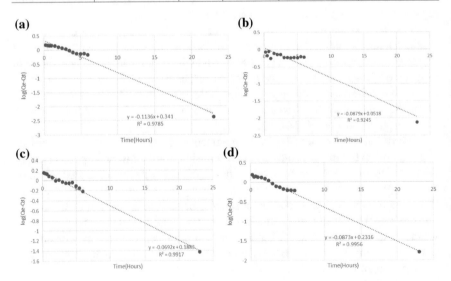

Fig. 3 First-order kinetic data for **a** PCP-MPAC; **b** P-MPAC; **c** C-300-MPAC; **d** C-700-MPAC

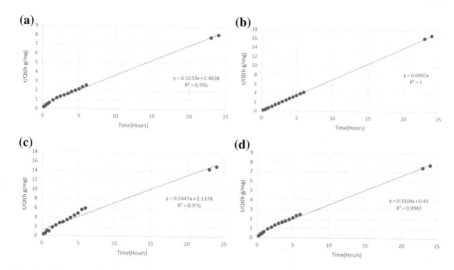

Fig. 4 Second-order kinetic data for **a** C-700-MPAC; **b** C-300-MPAC; **c** P-MPAC; **d** PCP-MPAC

PCP-MPAC shows the most superior removal and this could be attributed to quite a few factors. Mild carbonization plays an important role in proper chemical activation. Thus, sulphuric acid has a more enhanced effect of chemical activation since partly generated pores are already available. This also reduces the energy consumption in producing the activated carbon. Although it may seem that there are three steps in producing the activated carbon but certainly the amount of time spent activating at the higher temperature is essentially reduced making the process energy efficient in totality.

6 Conclusions

It can be safely concluded that MPs can act as suitable precursors for producing high-quality ACs. These ACs can be used to remove one of the most abundant pollutants of the textile industry namely methylene blue dye. Using MPs as a source for AC achieves the dual purpose of not only producing a substance capable of cleaning the waters of our planet but also helps reduce solid waste in the form of MPs that are either incinerated to produce gaseous pollutants or end up as landfill. Since MPs are a waste product of different beverage industries and mango beverages are available all through the year, this ensures that not only the raw material will be cheap but also available year round for continuous production of ACs. The batch adsorption studies proved that chemical activation along with thermal activation at suitable temperatures is necessary for proper performance of the adsorbent. High-temperature activation combined with low-temperature pre-carbonization proving essential to get the desired qualities in the AC.

References

Ajila, C. M., Bhat, S. G., Rao, U. J. S. P. (2007). Valuable components of raw and ripe peels from two Indian mango varieties. *Food Chemistry, 102*, 1006–1011.

FAOSTAT, FAO Statistical Database—Agriculture, http://apps.fao.org. 2005.

Lakshmi, U. R., Srivastava, V. C., Mall, I. D., & Lataye, D. H. (2009). Rice husk ash as an effective adsorbent: Evaluation of adsorptive characteristics for Indigo Carmine dye. *International Journal of Environmental Management, 90,* 710–720.

Lillo-Rodenas, M. A., Carzrola-Ameros, D., & Linares-Solano, A. (2003.) *Carbon, 265* (41).

López-Grimau, V., & Gutiérrez, M. C. (2006). Decolourisation of simulated reactive dye bath effluents by electrochemical oxidation assisted by UV light. *International Journal of Chemosphere Research, 62,* 106–112.

Pehlivan, E., Yanik, BH., Ahmetli, G., & Pehlivan, M. (2008). Equilibrium isotherm studies for the uptake of cadmium and lead ions onto sugar beet pulp. *Bioresource Technology, 99*, 3520–3527.

Shukla, S. R., & Pai, R. S. (2005). Removal of Pb(II) from solution using cellulose containing materials. *Journal of Chemical Technology and Biotechnology, 80,* 176–183.

Wu, J. S. B., Chen, H., & Fang, T. (1993). Mango juice. *Fruit juice processing technology* (pp. 620–655). Auburndale: Agric Science.

Xuan, Z. X., Tang, Y. R., Li, X. M., Liu, Y. H., & Luo, F. (2006). Study on the equilibrium, kinetics and isotherm of biosorption of lead ions onto pretreated chemically modified orange peel. *Biochemical Engineering Journal, 31,* 160–164.

Passive High-Rate Wastewater Treatment Using Upwelling Anaerobic Lagoons as an Alternative to Conventional Systems

Seema Sukhani and Hoysall Chanakya

Abstract Wastewater treatment processes employed today are many decades old and often do not follow emerging criteria of sustainable development promoting resource recovery as well as smaller environmental footprint while ensuring nutrient reuse and stakeholder involvement. Based on various performances reported and field observations of various poorly operating STPs in South India, it is proposed that an intermediate concept of anaerobic upwelling sludge bed reactor (USB) is proposed to operate with no O&M interventions like a septic tank but providing potential for methane and nutrient recovery at rates and HRTs as practised today, namely HRT < 24 h, loading rates 1–2 kg BOD/m^3/d, 70–80% BOD removal, 1–3 years of attention-free operation and moderate effluent turbidity. Some field scale evidence of such function at various locations is provided, and its proposed operation is discussed.

Keywords Sewage treatment · SDG-6 · Upwelling sludge bed reactor · Resource recovery · International society of waste management, air and water

1 Introduction

Developing countries face severe shortage of wastewater treatment infrastructure comprising of sewerage and collection systems, wastewater treatment and sludge disposal facilities. Lack of financial assistance and technical expertise has been some of the biggest factors for incapability to construct and operate conventional centralized STP. Absence of a reliable power supply and high costs of backup diesel power and frequent maintenance needs have been a major concern in rural and peri-urban areas in developing countries and hindered widespread deployment of STPs. Such factors also need to be considered before any technological development efforts are taken up (Verawaty et al. 2013).

S. Sukhani · H. Chanakya (✉)
Indian Institute of Science, Bangalore, India
e-mail: chanakya@iisc.ac.in

S. Sukhani
e-mail: seemasukhani@iisc.ac.in

© Springer Nature Singapore Pte Ltd. 2020
S. K. Ghosh et al. (eds.), *Recent Trends in Waste Water Treatment and Water Resource Management*, https://doi.org/10.1007/978-981-15-0706-9_23

Further, as various countries strive to provide compliance and capabilities for various sustainable development goals, SDG-6, especially in urbanizing locations, the absence of reliable power supply hinders the use of high-rate aerobic systems as well as power-based control systems for the STPs to be set up. Thus, there is a need to develop energy neutral or energy-independent STP systems that are also ecologically safe, socially acceptable, economically viable and self-sustaining in the immediate future while also being simple to operate (Luo et al. 2014). It is believed that when small village settlements those have rapidly grown and become urbanized, the high costs of land as well as the absence of large spaces for local treatment will drive them towards decentralized STPs, and these recently urbanized locations will be the first to face such problems of the absence of appropriate technology for such locations. In this paper, a review is attempted from the capabilities of various systems focussing on low-energy systems that are likely to be suitable to the emerging sustainability, people-led and multiple by-product-driven scenario of the future.

2   Conventional Passive and Low-Energy STPs

Constructed wetlands are one of the best-known passive systems and well established with regard to their performance. Their potential is highly dependent on microbial activity, retention times, variation in load, particulate content and climatic factors such as precipitation and evapotranspiration (El-Khateeb et al. 2009; Katsenovich et al. 2009). However, major limitations of system are poor nutrient (nitrogen and phosphorus removal), large areal footprint, propensity to clog during overloads and oxygen transfer limitations.

In urban and peri-urban areas, large footprint, which is essential for good effluent in constructed wetlands (Kivaisi 2001; Foladori et al. 2013), is difficult to acquire. Footprints can vary around 0.06kg BOD/m^2d (Zhang et al. 2009; Vymazal 2011). Nutrient recovery is also very low. In a study conducted by Vymazal in 2007, out of 40–55% of N removed (initial load 250 and 630 g Nm^{-2} $year^{-1}$) and ~40–60% of TP (initial load 45 and 75 g Pm^{-2} $year^{-1}$) removed, only 100–200 g Nm^{-2} $year^{-1}$ and 10–20 g Pm^{-2} $year^{-1}$ could be harvested through vegetation biomass. Important recoverable are being lost to environment to cause pollution on the one side and unsustainability on the other. In the emergent sustainability scenario, it is imperative that a significant fraction of the sewage nutrients are harvested and made amenable for recycling and reuse.

Aerobic systems such as activated sludge process are most commonly adopted treatment technology. As mentioned earlier, for a sustainable system, high treatment efficiency and low energy consumption are the key. In aerobic systems, energy demands are high due to operation of pumps, air blowers, mixers and separators. In general, 1–25 kWh/kg COD removed is required to operate various types of aerobic system depending on technology as compared to 0.25–13 kWh/kg COD removed for anaerobic systems (Amanatidou et al. 2016; Pitas et al. 2010; Yin 2010). In a conventional system, observed biomass production rate (Y_{obs}) usually ranges between 0.12 and 0.45 kg MLVSS kg COD removed-1 (Amanatidou et al. 2016).

Such excessive sludge will hinder sedimentation and create problems while mixing and aeration. Ultimately, a large part of sludge needs to be wasted, which again has significant economic and ecological footprint in terms of drying the sludge and disposing it in landfills. It must be noted that most of the aerobic systems of today are between 50 and 150 year old and now need alternatives in the light of emerging sustainability criteria of energy, C and environmental footprint as well as resource recovery criteria.

2.1 Anaerobic System

Anaerobic treatment systems generally have low operating costs and lower environmental footprint, and they also produce biogas, compost and many value-added products. However, for reasons of a slightly lower treatment efficiency and slower reaction rates and susceptibility to near-freezing temperatures, they have often been overlooked. Effluent from an anaerobic needs a small but definite aerobic treatment. Despite these 'stated' problems associated with anaerobic systems, they still are our best option for BOD removal. Though, as mentioned above the energy requirements are lower than all systems, potentially higher loading rates than aerobic systems, most planners are not generally in favour of the anaerobic systems for various reasons of skills and monitoring needs. The future capabilities or potential would be set by high-end processes such as UASB while seeking the O&M simplicity of a septic tank. Some of the studies conducted on UASB and septic tank in various parts of world are mentioned in Table 1.

2.2 Sludge Production

Biomass production rate (Y_{obs}) during anaerobic metabolism is usually between 0.03 and 0.15 kg MLVSS/kg COD removed. Moussavi et al. (2010) after operating septic tank for 355 days concluded that depth of total sludge produced and inert accumulated was still less than half volume of reactor. This illustrates that even in simple system like septic tank, there is no need to frequently remove sludge. It was observed by Lin and Yang (1991) that sludge with a concentration of 30–40 kg TS/m^3 exhibits the highest methanogenic-specific activity. Yet under anaerobic conditions and depending upon the organisms responsible, the Yobs of sludge can range between 0.3 and 0.5 kg MLVSS/kg COD removed. There are instances of 3–5 year uninterrupted operation of anaerobic systems without need for sludge removal. Older anaerobic sludges tend to become heavier and gradually decompose to be become a lot more

Table 1 Performance of anaerobic reactors across various processes and locations

Location	Type of wastewater	Volume (m³)	Temp °C	Influent conc. (mg/L) COD	BOD	HRT (h)	Removal efficiency (%) COD	BOD	Type	References
Columbia	Sewage	64	25	267	95	6–8 h	75–82	75–93	UASB	Kooijmans and Van velsen (1986), Lettinga et al. (1987)
Brazil	Sewage	120	19–28	627	357	4	74	78	UASB	Barbosa and Sant' Anna (1989)
India	Sewage	1200	20–30	563	214	6	74	75	UASB	Draaijer et al. (1992)
India	Sewage	6000	18–32	404	205	8	62–72	65–71	UASB	Haskoning (1996), Tare et al. (1997)
Finland	Synthetic blackwater	np	10–20	400–1900	NA	4.4d	90	np	UASB-septic tank	Luostarinen and Rintala (2005)
Netherlands	Domestic wastewater		12–20	2000–3000		6	51.5–60		UASB	Zhang et al. (2018)
Iran	Sewage (mix black and grey)	0.25	18–36	154–395	119–261	24	77%	85%	Septic tank	Moussavi et al. (2010)
Iran	Sewage (mix black and grey)	0.5	Dec-26	154–395	119–245	12	67	71	Septic tank	Moussavi et al. (2010)
Egypt	Concentrated sewage	0.575	13	3600		120	94		Upflow hybrid septic tank	Adhikari & Lohani (2019)

dense, and aerobic sludge is a lot more bulky and thus typical aerobic treatment systems require weekly and monthly management of sludge. Needless to say, aerobic sludges need a lot more careful treatment since they are quite odoriferous and bulky, thus more difficult to dewater and require monthly management of sludge.

2.3 Retention Time

The reaction rates are longer in typical septic tanks due to the absence of sludge recycling (maintaining a higher level of MLVSS), a gradual accumulation of inert solids and non-specific selection of sludge. Also, the time required to reach at a steady state is somewhat long taking 1–3 months. However, in a UASB system, these factors are overcome by maintaining a specific upflow velocity, which firstly helps retaining a high level of MLVSS that is constantly recycled, and this in turn tends to select a good settleable sludge with high microbial mass (lower EPS content). The retention period for same level of treatments is approximately three times lower achieving loading rates in the range of 3–8 kg $COD/m^3/d$. This is a lot higher than typical aerobic system that can handle only about 1–2 kg $COD/m^3/d$. From this, it is obvious that we need to engineer a system which can give us treatment efficiency and throughput of a UASB and energy independence/simplicity of a septic tank. This appears feasible if it is possible to hold on to a lot of heavy anaerobic sludge in the reactor and make it to be recirculated in a passive manner without external energy application. This invokes the concept of natural phenomenon of sludge upwelling if particles are allowed to mix with the sludge at a specific rate.

2.4 Sludge Properties and Activity

Under a typical mesophilic condition of 15–28 °C, the typical density of anaerobic sludge would be 1.05–1.10 g/m^3, and this imparts a propensity to settle rapidly. Being anaerobic, there is very little tendency to accumulate a large fraction of extracellular polysaccharides (EPS) very characteristic of aerobic sludges. When fed at a near maintenance rates, the settling sludge maintained would be primarily of viable bacteria. It is thus possible to create a process that selects predominantly well-flocculated sludge, thereby gradually enhancing its settling properties. This is typical in anaerobic sewage treatment systems giving a flocculated sludge in range of 20–30 ml/g of sludge. Although granular sludge has higher methanogenic activities with SVI 10–20 ml/g/d, they are useful for soluble COD and much less for sewage. The role granular sludge (e.g. UASB) in particulate COD removal before hydrolysis has been reported to be very limited. In terms of activity, 1.2 and 1.9 kg/kg VSS/d of methanogenic activity is appropriate to achieve a F/M ratio of 0.5 or even lower. It has been observed that in lower loading rates, diversity of micro-organisms is much more and they have capability to degrading diverse compounds. In anaerobic sludge,

VSS/SS ratio observed is 0.70–0.85 and wet cakes after centrifugation have 10% solids. This sludge can be used as a fertilizer by directly applying it to field.

3   Conclusion

Most wastewater systems being adopted and run today are very old and do not meet many sustainability criteria of Carbon, nutrient, energy and environmental footprint as well as societal inclusion. Meeting SDG obligations, especially to emerging urbanization, will require newer concepts of decentralized treatment with resource recovery. From various performance criteria found in literature, it is suggested that there is a need for modifying anaerobic reactors to provide the O&M and design simplicity of septic tanks while creating performance yardsticks closer to UASB processes.

The upwelling sludge bed reactor concept has been advanced to meet immediate application goals and performance criteria. Anaerobic system with one-third volumetric sludge, F/M ratio <0.5, 1.2–1.9 kg/kg VSS/d of methanogenic activity with functional MLVSS to be ~8–9 g/L, natural mixing through sludge upwelling is proposed. Based on field observations at HRT ~24 h, approximately 70–80% COD and 90% BOD removal occurs in such system. The proposed design and upwelling phenomenon are shown in Figs. 1 and 2, respectively. This system can be applied from a scale of 1MLD to a community level. This unit is a part of a sequential system and is discussed elsewhere.

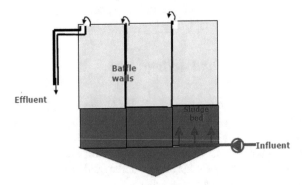

Fig. 1 Conceptual reactor design for passive sewage treatment

Fig. 2 Anaerobic waste-water treatment by sludge upwelling followed by *Rhodopseudomonas* stage (In Karnataka, India. Unpublished data)

References

Adhikari, J. R., & Lohani, S. P. (2019). Design, installation, operation and experimentation of septic tank – UASB wastewater treatment system. *Renewable Energy, 143*, 1406–1415. https://doi.org/10.1016/j.renene.2019.04.059.

Amanatidou, E., Samiotis, G., Trikoilidou, E., Tzelios, D., & Michailidis, A. (2016). Influence of wastewater treatment plants' operational conditions on activated sludge microbiological and morphological characteristics. *Environmental Technology, 37*(2), 265–278. https://doi.org/10.1080/09593330.2015.1068379.

Barbosa, R. A., & Sant'Anna, G. L. (1989). Treatment of raw domestic sewage in an UASB reactor. *Water Research, 23*(12), 1483–1490. https://doi.org/10.1016/0043-1354(89)90112-7.

Draaijer, H., Maas, J. A. W., Schaapman, J. E., & Khan, A. (1992). Performance of the 5 MLD UASB Reactor for Sewage Treatment at Kanpur, India. *Water Science and Technology, 25*(7), 123–133.

El-Khateeb, M. A., Al-Herrawy, A. Z., Kamel, M. M., & El-Gohary, F. A. (2009). Use of wetlands as post-treatment of anaerobically treated effluent. *Desalination, 245*(1–3), 50–59. https://doi.org/10.1016/J.DESAL.2008.01.071.

Foladori, P., Ruaben, J., & Ortigara, A. R. C. (2013). Recirculation or artificial aeration in vertical flow constructed wetlands: A comparative study for treating high load wastewater. *Bioresource Technology, 149*, 398–405. https://doi.org/10.1016/J.BIORTECH.2013.09.099.

Haskoning. (1996). *14 MLD UASB treatment plant in Mirzapur, India. Evaluation report on process performance*. Haskoning Consulting Engineers and Architects.

Katsenovich, Y. P., Hummel-Batista, A., Ravinet, A. J., & Miller, J. F. (2009). Performance evaluation of constructed wetlands in a tropical region. *Ecological Engineering, 35*(10), 1529–1537. https://doi.org/10.1016/J.ECOLENG.2009.07.003.

Kivaisi, A. K. (2001). The potential for constructed wetlands for wastewater treatment and reuse in developing countries: a review. *Ecological Engineering, 16*(4), 545–560. https://doi.org/10.1016/S0925-8574(00)00113-0.

Kooijmans, L. J., & van Velsen, E. M. (1986). Application of the UASB process for treatment of domestic sewage under subtropical conditions, the Cali case, In *Anaerobic Treatment: A grown up technology*. Papers of the IAWQ-NVA conference on Advanced Wastewater Treatment (Aquatech 1996), 423–426. Amsterdam, The Netherlands.

Lettinga, G., de Man, A., Grin, P., & Hulshoff Pol, L. (1987). Anaerobic wastewater treatment as an appropriate technology for developing countries. *Tribune Cebedeau, 40*(519), 21–32.

Lin, K., & Yang, Z. (1991). Technical review on the UASB process. *International Journal of Environmental Studies, 39*(3), 203–222. https://doi.org/10.1080/00207239108710695.

Luo, Y., Guo, W., Ngo, H. H., Nghiem, L. D., Hai, F. I., Zhang, J., … Wang, X. C. (2014). A review on the occurrence of micropollutants in the aquatic environment and their fate and removal during wastewater treatment. *Science of the Total Environment, 473–474*, 619–641. https://doi.org/10. 1016/J.SCITOTENV.2013.12.065.

Luostarinen, S. A., & Rintala, J. A. (2005). Anaerobic on-site treatment of black water and dairy parlour wastewater in UASB-septic tanks at low temperatures. *Water Research, 39*(2-3), 436–448.

Moussavi, G., Kazembeigi, F., & Farzadkia, M. (2010). Performance of a pilot scale up-flow septic tank for on-site decentralized treatment of residential wastewater. *Process Safety and Environmental Protection, 88*(1), 47–52. https://doi.org/10.1016/J.PSEP.2009.10.001.

Pitas, V., Fazekas, B., Banyai, Z., & Karpati, A. (2010). Energy efficiency of the municipal waste water treatment. *Journal of Biotechnology, 150*(150), 163–164. https://doi.org/10.1016/j.jbiotec. 2010.08.424.

Tare, V., Mansoor, A.M., & Jawed, M. (1997). Biomethanation in domestic and industrial waste treatment-an Indian scenario. In *Proceedings of the Eighth international Conference on Anaerobic Digestion*, May 25–29, 1997, 2, pp. 255–262. Sendai: Document of the IAWQ, Pergamon Press.

Verawaty, M., Tait, S., Pijuan, M., Yuan, Z., & Bond, P. L. (2013). Breakage and growth towards a stable aerobic granule size during the treatment of wastewater. *Water Research, 47*(14), 5338–5349. https://doi.org/10.1016/J.WATRES.2013.06.012.

Vymazal, J. (2007). Removal of nutrients in various types of constructed wetlands. *Science of the Total Environment, 380*(1–3), 48–65. https://doi.org/10.1016/J.SCITOTENV.2006.09.014.

Vymazal, J. (2011). Plants used in constructed wetlands with horizontal subsurface flow: A review. *Hydrobiologia, 674*(1), 133–156. https://doi.org/10.1007/s10750-011- 0738-9.

Yin, C.-Y. (2010). Emerging usage of plant-based coagulants for water and wastewater treatment. *Process Biochemistry, 45*(9), 1437–1444. https://doi.org/10.1016/J.PROCBIO.2010.05.030.

Zhang, D., Gersberg, R. M., & Keat, T. S. (2009). Constructed wetlands in China. *Ecological Engineering, 35*(10), 1367–1378. https://doi.org/10.1016/J.ECOLENG.2009.07.007.

Zhang, L., De Vrieze, J., Hendrickx, T. L. G., Wei, W., Temmink, H., Rijnaarts, H., & Zeeman, G. (2018). Anaerobic treatment of raw domestic wastewater in a UASB-digester at 10 °C and microbial community dynamics. *Chemical Engineering Journal, 334*, 2088–2097. https://doi. org/10.1016/j.cej.2017.11.073.